Tunnelling: Management by *design*

Alan Muir Wood

GEORGE GREEN LIBRARY OF
SCIENCE AND ENGINEERING

London and New York

First published 2000 by E & FN Spon
11 New Fetter Lane, London EC4P 4EE

Simultaneously published in the USA and Canada
by E & FN Spon, an imprint of Routledge
29 West 35th Street, New York, NY 10001

E & FN Spon is an imprint of the Taylor & Francis Group

© 2000 Alan Muir Wood

Typeset in Sabon by Florence Production, Stoodleigh, Devon.
Printed and bound in Great Britain by Biddles,
Guildford and King's Lynn

British Library Cataloguing in Publication Data

A catalogue record for this book is available from the British Library

Library of Congress Cataloging-in-Publication Data
Muir Wood, A. M. (Alan Marshall)
  Tunnelling: management by design / Alan Muir Wood.
    p. cm. 1005268447
  Includes bibliographical references and index.
  ISBN 0-419-23200-1 (hb : alk. paper)
  1. Tunneling. 2. Tunnels–Design. I. Title.
  TA805 .M85 2000
  624.1´93–dc21

ISBN 0–419–23200–1   99-047534

Contents

Preface

Every author will claim that his or her book will cast light on aspects of the chosen topic not previously illuminated by his (or her – to be understood throughout) precursors. Where then are the dark unexplored recesses of the underground world that justify the promised light at the end of the tunnel of this present account?

The immediate spur to writing this book is that the author has lived and worked through a period of revolutionary change in tunnelling, with several components:

- change from traditional craft to technological art;
- spectacular advances in site investigation techniques and in geotechnical analysis;
- great strides in technological development in all aspects of tunnel construction;
- emphasis on the teachable elements of science applied to tunnelling;
- recognition of the interplay of opposites: opportunity and risk, in the development of tunnelling strategies;
- institutional recognition of tunnelling as a specific branch of engineering.

But costly mistakes – possibly costing more than the original estimate of the project – are now more common occurrences, usually of a foreseeable and preventable nature. Overall, therefore, the industry is nowhere near optimum potential, to the frustration of those who work in it, the wasting of personal effort, the thwarting of the objectives of the promoters of projects who, in the most egregious failures, have themselves through lack of understanding established conditions unconducive to success. Where lawyers earn far more from the failure of projects than do the most skilled engineers from success, clearly there are fundamental systemic faults.

From the surface, no single explanation for this contradictory situation is apparent but deeper digging indicates a common set of system failures.

A primary purpose of this book is therefore to see tunnelling as a system and to develop principles for success based on effective understanding and operation of the system of interplay of specific tunnelling skills.

The views expressed in this book are the author's own but he acknowledges his debt to his many immediate colleagues in Halcrow and to so many tunnellers and others around the world for several of the thoughts which have prompted the book. The author accepts full responsibility for any misunderstanding. The account may be criticised as unduly orientated towards the British and European examples. This is justified by selecting examples for which the circumstances are generally familiar; lessons learned may well have more universal application. The Channel Tunnel, for example, provides many examples of meritorious engineering with less meritorious management. An exemplary account of its engineering geology (Harris *et al.* 1996) attracts numerous references on account of its depth and breadth.

Notation

Against each symbol is a brief definition and a reference to the most appropriate Section or Appendix of the book for further explanation.

a	major semi-diameter of ellipse	9.3
a	tunnel radius	App. 5A
A	cross-sectional area of tunnel	5.3
b	minor semi-diameter of ellipse	9.3
c	cohesion	5.1.1
c_a	concentration of gas in air	8.3
c_u	undrained shear strength of soil	5.1.1
c_v	coefficient of consolidation	5.1.1
c_w	concentration of gas dissolved in water	8.3
C	construction hazard	6.1.3
C	circumference	9.3
C_p	velocity of compressive wave	4.1.2
C_s	velocity of shear wave	4.1.2
d_{10}	size of tenth smallest fraction	6.3
D	design of project hazard	6.1.3
E	Young's modulus for lining	App. 5A
E_c	Young's modulus for ground	App. 5A
F	flux rate for transfer of gas between air and water	8.3
h	depth of tunnel axis	App. 5G
h	depth of compensation grouting	5.3
H	height of column of rock supported by tunnel	App. 5D
H	head difference	App. 5G
H	geological hazard	6.1.3
H	Henry's constant	8.3
i	distance of point of inflexion from axis	5.3
I	second moment of area	App. 5A
k	hydraulic permeability	App. 5G
k_h	hydraulic permeability (horizontal)	App. 5G
k_v	hydraulic permeability (vertical)	App. 5G

K_0	earth pressure coefficient at rest	5.1.1
K_p	coefficient of passive pressure	5.2.1
K_p'	coefficient of effective passive pressure	App. 5B
m	radial spacing of rock bolts at rock face	App. 5D
m_v	coefficient of volume compressibility	5.1.1
M	bending moment in lining per unit length of tunnel	App. 5A
M_{max}	maximum value of M	App. 5A
n	longitudinal spacing of rock bolts at rock face	App. 5D
n	period of years	3.2
N	circumferential load in lining per unit length	App. 5A
N_s	stability ratio	5.1.1
p	V_s/A	5.3
p	probability	2.7
p_i	internal support pressure	5.1.1
q	surface surcharge pressure	5.1.1
q	flow per unit area	App. 5G
q_u	unconfined compressive strength	5.1.1
Q	weight of explosive charge	6.3
Q_0	inflow (outflow) per unit length of tunnel	App. 5D
Q_p	inflow per unit length of probe-hole	App. 5G
r	rate of interest	3.2
r	radial coordinate	App. 5A
r_0	radius of tunnel	App. 5A
R	geological risk	6.1.3
R_a	radius of curvature of ellipse on major axis	9.3
R_b	radius of curvature of ellipse on minor axis	9.3
R_c	competence ratio (q_u/σ_0)	5.1.1
R_s	compressibility factor	App. 5A
T	tension in rock-bolt	App. 5D
u	radial convergence	5.1.1
u_a	radial convergence at radius a	5.1.1
u_*	shear velocity	App. 5G
$\hat{u}$	maximum value of u	App. 5A
V_s	area of surface settlement trough	5.3
V_t	ground loss into tunnel per unit length	5.3
w	surface settlement	5.3
w_{max}	maximum value of surface settlement	5.3
y	coordinate along axis of tunnel	5.3
z	vertical coordinate	5.3
z_0	depth from surface to tunnel axis	5.1.1
Δr	change in radius, r	9.3
ΔC	change in circumference, C	9.3
γ	unit weight of soil	5.1.1

γ_w	unit weight of water	App. 5A
ϵ	strain	App. 5B
ϵ_r	radial strain	App. 5B
ϵ_y	axial strain	App. 5B
ϵ_θ	circumferential strain	App. 5B
θ	angular coordinate	App. 5B
λ	radial stress parameter	5.1.1
λ	coefficient of ground reaction	App. 5A
ν	Poisson's ratio	App. 5A
ρ	density	4.1.2
σ	compressive stress	5.1.1
σ_a	stress at radius a	5.1.1
σ_h	horizontal stress	5.1.1
σ_n	normal stress	5.1.1
σ_r	radial stress	5.1.1
σ_v	vertical stress	5.1.1
σ_x	bending stress	9.3
σ_y	longitudinal stress	App. 5B
σ_θ	circumferential stress	5.1.1
σ_0	initial stress in ground and far-field stress	5.1.1
τ	shear stress	5.1.1
$\tau_{r\theta}$	shear stress in plane r,θ	App. 5A
ϕ	angle of friction	5.1.1
ϕ	Airy function	App. 5A
ϕ	equipotential line	App. 5G
ψ	flow line	App. 5G

Introduction

Tunnelling, along with much civil engineering in Britain, particularly of projects of high potential risk, has suffered during the late 1980s and 1990s from the unsuccessful experiment – doubtless influenced by the prevailing politics of the time – to apply crude free market principles to the procurement of projects in fragmented elements, each element at least cost, centrally administered but not integrated. To many, the paradox has not been lost that a commercially motivated doctrine applied in an inappropriate manner to an essentially professional field reserved its hardest blows for the commercial interests themselves, as 'management' and 'engineering' were condemned to part company.

The jargon of 'systems' is unfamiliar to tunnelling. In consequence, in this book, a familiar term 'design', usually used in construction in far too narrow a sense, is used in the broad sense in which it is applied, for example, in manufacturing engineering (Chaplin 1989). Where the term 'design' is used in this comprehensive sense, as defined in Chapter 2, it is printed throughout as '*design*' to avoid confusion with the narrow sense of 'scheme design' and 'element design' traditional to tunnelling. One particular objective of this book is the emphasis on the essential interaction between product design (the design of the finished project and of its operation) and process design (the design of construction and its means) for success. The dominant role of *design* in this broad sense is discussed in Chapter 2.

The book is thus largely concerned with the features of operation of the *design* process to allow the development and execution of the optimal scheme by the interactions, often iterative, between the several contributors. Along the way, examples, many within personal experience, illustrate success and failure, and underscore the benefits of *design* to all concerned, not least the 'clients' and their financial supporters. There are those who maintain that *design* equates to engineering and that there is merit in emphasising this correspondence. The author has much sympathy with this view but 'engineering' already has too many connotations. The main virtue in *design* is that the engineer will be working with other disciplines who will be able to apply the same term *design* for their combined, and for their separate, activities. The term will therefore bond and not divide the team.

What is so special about tunnelling? Each particular property is shared to some degree with other forms of construction; many of the principles discussed have far wider application. Tunnelling may however be characterised by these features:

- extreme dependence on the ground, the interpretation of its characteristics in terms of risk;
- high degree of interdependence between planning and project design, arising from provisions for containing the ground and excluding ground-water;
- domination of the methods of construction on the design of the project;
- effect of restrictions of access on logistics, particularly in dealing with construction problems;
- interdependence of a long chain of control between the intention and its execution.

To use current jargon, the broad claim is the attempt to provide the first holistic account of tunnelling. Much has been written on design – conceived as analysis in determining the geometry and constituent elements of the finished tunnel, somewhat less on techniques of construction. There are also many accounts of underground planning from the last 20 years, a few monographs on site investigation for tunnels and briefer accounts of the elements of the overall process. Many Papers and Conference Proceedings describe specific case histories but few attempt to trace misfortunes to their fundamental causes and fewer yet attempt to provide analysis of the explanations for success or to provide clues as to how to avoid unfortunate repetition in the future. A certain amount has been written by tunnellers on the contractual and management aspects for success but this seems largely to have passed by those who set standards from the top-down style of much that passes for engineering management.

Successful tunnelling requires the blending of many skills, the acquisition of experience and judgement, and the transmission of the benefit of this experience to newcomers into the underworld. Much benefit results from encouragement to exchange views and to compare experiences. This is achieved at several levels, including the learned society activities of professional and technical bodies, at an international level by the activities – and with the encouragement of – the International Tunnelling Association (ITA), the International Society for Rock Mechanics (ISRM), the International Society for Soil Mechanics and Geotechnical Engineering (ISSMGE) and their specialist groups, also the Technical Committee on Tunnelling of PIARC (Permanent International Association of Road

Congresses). The most valuable contributions demonstrate the degree to which expectations based on analysis and experiment correspond to experience in projects, particularly where the account includes statements as to what has been learned. Certain classes of engineer appear to avoid exposing their approaches to their work, preferring instead to spread a deliberate air of mystery around their work as an esoteric art confined to initiates of a particular doctrine. This is nonsense and their own degree of success suffers in consequence. It is noticeable, for example, that sprayed concrete lining (SCL) tunnelling in soft ground (see Chapter 1) has profited immeasurably from the recent open comparison between theory and practice arising from many participants of its application to British clays.

The principal players in a tunnelling project may be imagined to be assigned as the members of an orchestra. Each needs to be able to master his own instrument, each needs to have a good ear for the contributions of others in order to be able to engage in the counterpoint of dialogue. The conductor, the leader of the project, needs to understand how to blend the contributions by the players, requiring an appreciation of the range of pitch and tonalities – the specific element – of each instrument. Too often, the tunnelling players are each following unrelated scores, with the conductor confined to the role of the orchestral administrator, without insight into the essence of the enterprise, the manager without understanding of what is managed. No wonder if the result is too frequently cacophonous.

This book is directed towards the interests and needs of each member of the tunnelling orchestra and of its impresarios, the Owners, to assist each in playing his own part and to become better attuned to the contributions of others. The objective is to help not only those engaged in the direct functions of achieving the underground project but also those in associated functions such as physical planning and transport economics, particularly those concerned in financing and management from a distance, who play vital parts in achieving the optimal project. Furthermore, the key messages on the management of tunnelling projects should have a wider audience, not only because of common features of different types of major project but also to influence the 'decision makers' who far too often have made the wrong decisions of policy, incapable of retrospective correction, through failure to understand the criteria for success. These may include the Development Banks, Government Ministers, Local Government officials, and both public and private owners of Utilities. The hope is that there may be some influence towards steering an initial course so that those who come to put policy into action do not need to expend unnecessary energies on attempts partially to mitigate consequences of original misdirection. The fundamental criteria for success have a broad application to any project with a significant element of natural risk.

At the central point of initial decision-making, the 'conductor' is familiar with tunnel design and construction, also with the operational and financial objectives. During the early stages, particularly before commitment to the nature of the project, or even whether a tunnelled option may be preferred, the functions of the 'conductor' may be undertaken in whole or in part by a 'surrogate client' or 'surrogate operator' who brings the requisite understanding and who represents the client (Owner, promoter or operator) in commissioning whatever studies and investigations may be expedient to reach a preliminary view on the options and to advise on the formation of the team who will have the duty to carry forward in an orderly process the project *design*. Too often, the first professional recourse by the client is to a lawyer who sets out to erect the barricades for protection against contractual conflict, casting other participants as antagonists to be attached to contractual chains and to be exposed to legal minefields. This is the worst possible starting point for a project which, for success, essentially demands much cooperation from the contributing Parties.

Many skills need to be harnessed for the successful tunnelling project. Whether or not the appropriate skills are recruited depends on the management of the project. Here we encounter the first set of problems.

All civil engineering projects have a purpose beyond civil engineering. Tunnelling is no exception but occupies a rather esoteric position in that the ultimate objective is not, as a general rule, directly associated with an underground solution. The extension of a metro system or certain forms of hydropower are exceptions to this general rule. Those who commission projects of these latter types may be expected to have greater familiarity with the criteria for success of underground projects than those, for example, who operate main-line railways or water supply who only venture deep below the surface once in a generation. Whatever the purpose, the total management of the project will determine whether the potential for success survives the initial phase.

Much, possibly too much, has recently been written on project management. The engineer concerned with underground projects must be familiar with the elements of good management, of special concern to his type of project, including the available tools based on information technology or IT, and will find it advantageous, but not necessarily for clear thinking, to be familiar with the jargon and the acronyms – but at all costs must avoid the dangers of lapsing into management-speak. The essentials of project management start from an understanding of the qualities of leadership, and the criteria for overall project optimisation. Management by remote direction is fatal for tunnelling; some of the fatalities are described in this book. Tunnelling, as with any complex activity in a previously unexplored environment, contains elements of uncertainty that need to be understood and controlled. Optimisation in consequence entails the control of risk and the counterpart of exploiting opportunities for innovation. A primary essential of the manager is therefore to understand the features of the tunnelling process, to influence the overall strategy from the moment of first recognition of the possibility of a tunnelling solution through to the operation of the finished project. Chapter 7 describes the most essential functions of the manager and, for those contemplating the commissioning of a possible tunnel, this may be the point of entry, leading into Chapter 3 which describes the process of project definition, from the outset, when the project is no more than a gleam in the eye or a hypothetical solution to a previously unsolved problem. Chapter 5 follows the evolution of the project, identifying the principal features which should influence decisions at each stage, to avoid later untoward consequences.

The scheme of project management has no unique structure. The principles need to be respected within the evident requirement for compatibility with the organisation of the Owner (or Commissioner) of the Project. Good project management may readily accommodate the several forms of Private and Public Finance of Projects. It is nevertheless recognised that the rigidity of the rules of certain Public – and some Private – bodies needs to be relaxed. The fragmentation of projects by the separate commissioning of different aspects of the work must also be abandoned since such practices inhibit the essential interactive features of *design*. Can the Owner afford to engage people at the initial stages with operational knowledge and the appropriate skills to interact with those who are to define the tunnelling option? If not, then he cannot afford to consider a tunnelled option.

Those who continue to visualise tunnels as examples of linear procedures may find a certain logic in the ordering of Chapters 3, 4, 5 and 6, titled respectively: Planning; Studies and Investigations; Design of the Tunnel Project; Design of Construction. This is partly an illusion, however. Constant reminders are provided of the fact that this arrangement is little more than a convenience for reference; the important feature is that these apparent phases should be viewed as continuous processes with many cross-connecting functions, the essence of the holistic approach. Following another line of thought, the different types of tunnel for different purposes might have been developed coherently through planning to operation. Traces of this cross-dimension will be found in elements of each chapter.

The interlinkage of activities may be illustrated by consideration of risk, defined in Chaper 2. The principal technical elements of uncertainty of a tunnelling project derive from the ground. The initial route planning may introduce more or less features of uncertainty, which will need investigation or control or at least strict definition in relation to possibilities of construction. Tunnel design is dominated by the ground; so too is construction. Different forms of construction will be vulnerable to different aspects of uncertainty. Hence optimisation is a continuing activity, taking account of all such considerations. While the perception and anticipation of risk is fundamental to successful *design*, it should be understood that risk in tunnelling seldom lends itself to statistical calculation but is more of a pragmatic Bayesian nature (whereby assessments of future risk may be constantly updated from data compiled from immediate past experience), for several reasons.

Firstly, geological variability is so wide that the degree of uncertainty of the ground can usually only be expressed between limits, progressively narrowed as the project develops. For example, the precise nature of possible water inflows and the positions of important ground interfaces may be sufficiently unpredictable to demand a form of observational approach (see Chapter 2).

Secondly, the risks are consequences of interactions of a site- and project-specific nature. Even, therefore, if averaged statistics existed of incidents with particular combinations of the several factors, their application to a specific project would have little significance. The most significant factor will be the degree of awareness of the possible consequences of the particular combination of circumstances and their anticipation.

Thirdly, as every tunneller will know, there is a prevailing ethos of every project, partially derived from the degree of shared interests in the definitions for success between the parties concerned. This ethos may profoundly influence the attitude to risk and the appropriate preparedness. This is the social dimension of tunnelling, a source of light or of darkness.

A structure for systematically listing the factors contributing to risk should nevertheless be devised. How to circumscribe risk and how to apportion responsibility for control and management feature throughout this book. Risk as a main cause of increased cost and delay is cultivated by absence of continuity of purpose in project development, the associated fragmentation of responsibility, the failure early in the life of a project to address risk and its attempted evasion by transmission to other parties further down the line, involved with only a part of the project, who have remote prospects of definition, investigation or control. This book is concerned in developing the culture for enhancing opportunities, for the encouragement of good practices to permit the potential for successful projects and for the full realisation of satisfying project experiences.

While certain aspects of the relationships between the parties to a tunnelling project need to respect commercial criteria, in essence success depends upon the mutual recognition of the professional standards of the participants. The nature of *design* emphasises that these are paramount, between promoter, engineer, contractor and specialist. Recognition of this mutual respect is more highly cultivated in European countries other than Britain than in Britain and the USA, where the engineers have less control over the practices in procuring the contributory elements of tunnelling projects. The consequent loss is shared by all parties, with greater risk of loss and less innovation in practice. There are solutions to this sub-optimal culture, whose exposure and amelioration are the principal purposes of this book.

All of these aspects have a bearing upon the education and training of engineers. Tunnelling provides an excellent example of the need to combine depths – the understanding and expertise in several quite different disciplines – with breadth, the ability to fit all these elements together. Some suggest that technical design in the 'sharp' parts of engineering is quite a different activity from managing construction. This is a dangerous myth which promotes the separation between the two activities of design and management. The import of this book is that the two activities merge imperceptively and that it is the ability to achieve this merger which is a principal criterion for success.

Chapter 1

Background to modern tunnelling

If we can think of one bit of time which cannot be sub-divided into even the smallest instantaneous moments, that alone is what we can call the 'present'. And this time flies so quickly from the future to the past that it is an interval with no duration.

St Augustine, Confessions XI.xv (20), tr. H. Chadwick.

1.1 Introduction

The reader will find here no attempt to provide a comprehensive history of tunnelling. The objectives are more modest and more targeted, namely, to set the scene from the earliest known tunnelling, to review developments which have contributed to present-day understanding of the criteria for success, with references to exemplary projects.

On the negative side, the objective is to attempt to explain historical factors contributing to the current lack of general appreciation for these same criteria for success. In so doing, it is necessary to understand how circumstances have changed with time and with the increased complexity of the tunnelling operation. The many contributory factors include the increasingly specific technical requirements to satisfy the objectives of an underground project, developments in the techniques and the means of tunnelling, also in the associated improved capability to explore the ground, to measure and model its characteristics.

This account is in consequence deliberately selective, using particular examples, from the several periods of development of tunnelling, as landmarks, generally to emphasise the varying degrees of intuition, craft-skill and technology which have characterised the several periods into which this chapter is divided. There is no clearly defined boundary between tunnelling history and current developments. The most important features of current practice rest upon concepts which have been developing over many years. The sub-division between past and present is therefore one of convenience in telling a reasonably continuous tale.

A broad survey of tunnelling rapidly reveals how tunnelling has progressively been challenged to penetrate ground of increasing degrees of inherent difficulty. With time, in consequence, the relationship between the main parties concerned with each tunnel has assumed greater importance. A contrary more recent development, during a period in which inappropriate contract relationships have been increasingly imposed, adds to a mismatch between what these relationships should be and what in fact they have been. A pervading objective of this book is to help to guide towards good practice for the future, in the management aspects of engineering as much as the technology, the two being indivisible in good practice. A historical perspective provides a guide, with the view that knowledge of the point of departure is necessary in order to chart the course to our destination. For general accounts of the history of tunnelling, the reader may turn to Sandström (1963), Széchy (1970) and Harding (1981) although it will be found that more recent information causes the account below to make some modification to these texts. Background reading into earlier times may start from Singer *et al.* (1954).

1.2 Tunnelling in antiquity

Since tunnelling, especially tunnelling for mining minerals and for water supply, precedes recorded history, we can only conjecture as to the thought processes of those in the earliest times who discovered that tunnelling provided the solution to perceived requirements. We need to respect the intelligence of these early tunnellers of 3000 years ago and more for the magnitude of their achievements, whose art depended entirely on trial-and-error (heuristics) to learn what could be done and how best to achieve results, with neither the benefit of underlying technology nor with access to specialised tools. All ancient civilisations have left behind examples of tunnelling at varying scales.

The earliest tunnels would have been modelled on natural caves, selecting rock types capable of penetration with crude tools, and having innate strength and tight joint structure conducive to a natural stable form without need for artificial support. Much of the tunnelling for mining was initially of this nature following lodes with near-surface exposures. Early tunnels reflect the change of social priorities with time. Whereas at the present day a tunnel has one or more special purposes as conveyor of people or substance, storage or protected housing, many of the most spectacular tunnels were for ceremonial and religious purposes. It is remarkable that the first known sub-aqueous tunnel was built in about 2000 BC to connect the royal palace of Queen Semiramis to the Temple of Jove (or his equivalent) beneath the River Euphrates. The tunnel was about 1 km long, had a section of 3.6 m × 4.5 m, and was lined in brickwork with a vaulted arch set in bituminous mortar. The river was diverted to permit cut-and-cover construction. While this may not constitute tunnelling in the purest sense, nevertheless the sophistication of the techniques point to a pre-existing skill and assurance in undertaking such works.

It also deserves record that, for city drainage systems during the Akkadian supremacy (BC 2800–2200) of Mesopotamia, vaulted sewers were built in baked brick, with inspection chambers and rodding eyes, lined in bitumen. To wander a little further from our scene, there were bath-rooms paved in bitumen-faced brick, also closets with raised pedestals with the occasional refinement of a shaped bitumen seat (Singer *et al.* 1954). Rings of brickwork in vaults for wide spans were inclined to avoid the need for centring.

Tombs and temples provide other examples of tunnelling. The Royal Tombs at Thebes in Egypt and at Ur in Mesopotamia date from around 2500 BC; Abu Simbel in Egypt dating from around 1250 BC is another such example. These tunnels were in limestone, requiring high standards of craftsmanship in their form but presenting no problems of stability. Another later example, mentioned by Sandström (1963) among many examples in India, concerns the caves of Ellora, near Bombay, excavated between 200 and 600 AD, tunnels aggregating to a length of more than 9 km, cut by chisel in fine-grained igneous rock.

A celebrated, but by no means unique, water supply tunnel of the ancient world is the Siloam tunnel (also known as Hezekiah's tunnel), mentioned at 2 Kings 20:20, at 2 Chronicles 32:30 and elsewhere in the Old and New Testament of the Bible. The oldest part of the city of Jerusalem, dating from 2000 BC or earlier had a source of water from the Gihon spring on the Ophel ridge to the east of the city. The name Gihon in Hebrew (and its Arabic form, Umm al-Darah) relates to its gushing siphonic nature, with high flows for a duration of 30 minutes or more at intervals of 4–10 h (depending on the season and the source of information). As early as about 1800 BC, a short length of tunnel connected the spring to a well shaft nearly 30 m deep within the city wall. In around 950 BC Solomon connected the spring by an open channel outside the city wall to the pool of Siloam within the wall. Under threat of siege by the Assyrians under Sennacherib, King Hezekiah secured the water supply by resiting the pool of Siloam and feeding it through a tunnel 533 m long, following a sinuous route about 60% longer than the direct length from spring to outlet portal. The tunnel is roughly 1 m wide and 2 m high, increasing to a considerably greater height near the portal, presumably as a result of correcting an error in level. The reasons for the devious course of the tunnel have long been the cause of speculation. Most probably the explanation includes geological influences (natural fissures in the rock) combined with the avoidance of areas of royal tombs. The tunnel was advanced from both ends and a sharp Z-bend near mid-length probably marks the point of break-through. A stone memorial tablet found in 1880 provides, in one of the oldest example of cursive Hebrew script, a brief account of the tunnel and the occasion of break-through, which reads in translation: 'While the hewers yielded the axe, each man towards his fellow, and while there were still three cubits to be hewn, there was heard a man's voice calling to his fellow, for there was a crack in the rock on the right and on the left.'

The *qanats* of the Middle East, an art which survived over the centuries with little change, required a greater degree of understanding of the need for an internal lining to provide support and a watertight invert. Joseph Needham (1971) finds evidence of similar tunnels in China as early as 280 BC and conjectures as to whether the art may have passed from China to the Middle East or *vice versa*. The *qanats* demanded a degree of conceptual thinking since, as illustrated in Figure 1.1, water was inter-cepted from springs beneath the surface of the ground, derived from the face of a range of hills, the tunnels serving as conduits to wells, or to surface discharge as ground levels permitted, to satisfy requirements for irrigation. The *qanats* were constructed from generally vertical shafts at centres of about 150 feet (45 m), their courses marked by rings of spoil around each shaft. Intermediate access for water could be by way of stepped inclined shafts. The water supply to several cities of Asia Minor used ancient canals, partially tunnelled, of considerable length. Arbil in North Iraq, for example, continues to rely upon such a canal of Sennacherib, 20 km long with a tunnel whose ashlar entrance portal is 2.7 m wide, with floor and walls faced in stone slabs.

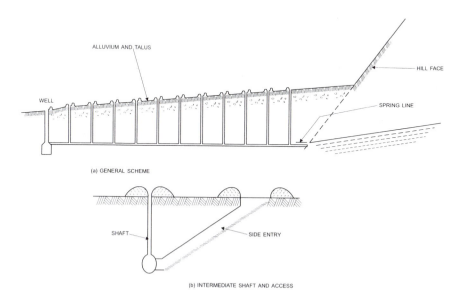

Figure 1.1 Qanat.

The Greeks and Romans were familiar with the use of tunnelling to undermine ('sap') the walls of defended cities, and in the appropriate countermeasures. Many tunnels were also constructed as part of the aqueducts supplying water to cities, such as that for Athens, remaining in use to the present day.

The Greek water supply tunnel on the island of Samos (Plichon 1974) highly regarded by Herodotas, merits a brief description. The tunnel, conceived by Eupalinos and constructed in 525 BC, provided water from a spring to the town of Samos. The tunnel has the form of an approximately square heading 1.8 m × 1.8 m, the roof generally provided by a stratum of competent rock. In order to maintain a continuous gradient for the water conduit, formed in tiles, a 0.8 m wide trench was sunk from the heading to depths between 1.5 m and 8 m, a series of stone slabs above the conduit supporting backfill in the trench of spoil from the tunnel.

The Romans have left such fine examples of masonry vaulting that it is no surprise that, for ground capable of standing unsupported over short lengths for a short time, they have demonstrated a capability for building arched tunnels of appreciable span. Their permanent stability depends in the same manner on transmission of load around the arch approximately transverse to the joints between voussoirs. Drainage tunnels mentioned by Livy include the 2210 m (1500 Roman *passi*) long tunnel to drain Lake Albanus. The tunnel, about 1.5 m wide by 2.5 m high, was constructed in hard volcanic rock in the year 359 BC, using working shafts at intervals of about 45 m. The most notorious Roman drainage project is that of the Lake Fucinus (now Celano) Emissarium, about 5500 m long, 2.5 m wide and 5 m high, which was driven from working shafts at intervals of about 35 m (a Roman *actus*), vertical and inclined, with ashlar used for lining in unstable ground, built between 41–51 AD. Suetonius describes the first unsuccessful opening ceremony after which the tunnel had to be regraded, the deferred ceremony leading to flooding of the celebratory banquet alongside the tail-race canal.

Road tunnels through spurs of the Apennines around the Bay of Naples were constructed at Cumae and at Puteoli, by Octavian (later Augustus) respectively about 1000m and 700 m long, 3 m wide and 3.2 m high. Several tunnels of the ancient world are mentioned by Livy, Pliny the elder, Herodotus, Suetonius and Vitruvius, with many of the references given by Humphrey *et al.* (1998).

1.3 Development of rationale

The breakdown in urban society with the collapse of the Roman empire caused the general absence of demand for tunnelling in Europe for several hundred years, although small-scale mining continued, using the traditional methods described by Agricola (Georg Bauer) in 1556. Széchy (1970) mentions a 5.6 km long drainage tunnel of the Biber mine in Hungary, started around 1400 as an outstanding example of the time. Techniques of timbered support and the methods of excavation were largely derived from mining practice. The construction of the Col di Tenda tunnel on the road between Nice and Genoa was started by Anne of Lusignan in 1450 but later abandoned (Harding 1981) then restarted in 1782 and once again abandoned in 1794 at a length of 2500 m.

Leonardo da Vinci made numerous proposals, which came to nothing, for navigation canals, including a connection between the rivers Garonne and Aude, subsequently constructed as the Languedoc canal by Pierre-Paul Riquet (Sandström 1963), including the Malpas tunnel (157 m × 6.7 m high × 8.2 m wide) completed in 1681, using gunpowder to blast the rock, but only lined some years later. The celebrated French engineer, Vauban, was also associated with this project.

Sandström contrasts the subsequent practices in canal building in Britain and in Europe, the former by privately financed canal companies at least cost, the latter, financed by the state, having more substantial dimensions. The first English canal tunnel was that of James Brindley on the Bridgewater canal directly entering Worsley coal mine, opened in 1761 and subsequently extended. This marked the beginning of the canal age, with numerous tunnels on the extending canal network. Brindley's next achievement was the Harecastle tunnel, 2630 m long, 4.3 m high and 2.7 m wide, on the summit of the Grand Trunk (later the Trent and Mersey) Canal, damaged by mining subsidence and replaced by a parallel tunnel constructed by Telford. The canal tunnels constructed by Outram, Jessop, Rennie, Telford and others set traditional standards for tunnel construction in rock and in weaker ground, using the English method described below.

By the seventeenth century, designers of the earliest canal tunnels in France were beginning to apply principles of graphical statics to the criteria for constructing stable arches. The design of timber support was also being developed from the experience in mines. Salt mines at Wielicka near Cracow in Poland, for example, contain massive timbering from the seventeenth century, protected by the saline atmosphere and providing evidence of a well-developed art.

Temporary timber support for tunnelling evolved in different parts of Europe, principally for canal and the early railway tunnels, attracting regional designations and having characteristics adapted to the particular conditions of the locality. Some of the better known systems, illustrated for example by Sandström (1963) and Harding (1981), had particular characteristics.

The *German system* provided a series of box headings within which the successive sections of the side walls of the tunnel were built from the footings upwards, thus a forerunner of the system of multiple drifts. The method depends on the central dumpling being able to resist without excessive movement pressure transmitted from the side walls, in providing support to the top 'key' heading prior to completion of the arch and to ensuring stability while the invert arch is extended in sections.

The *Belgian system* started from the construction of a top heading, propped approximately to the level of the springing of the arch for a horseshoe tunnel. This heading was then extended to each side to permit construction of the upper part of the arch, which was extended by underpinning, working from side headings. The system was only practicable where rock loads were not heavy.

The *English system* also started from a central top heading which allowed two timber crown bars to be hoisted into place, the rear ends supported on a completed length of lining, the forward ends propped within the central heading. Development of the heading then allowed additional bars to be erected around the perimeter of the face with boards between each pair to exclude the ground. The system is economical in timber, permits construction of the arch of the tunnel in full-face excavation, and is tolerant of a wide variety of ground conditions, but depends on relatively low ground pressures.

The *Austrian system* required a strongly constructed central bottom heading upon which a crown heading was constructed. The timbering for full-face excavation was then heavily braced against the central headings, with longitudinal poling boards built on timber bars carried on each frame of timbering. As the lining advanced, so was the timbering propped against each length to maintain stability. The method was capable of withstanding high ground pressures but was particularly extravagant in the demand for timber.

In the absence of other than primitive means for foreseeing the nature of the ground ahead of the advancing tunnel, there were frequent surprises. Linings were in masonry or brickwork depending on local availability of supply. Some of the early railroad tunnels in North America were lined in timber, with long, shaped voussoirs supported on vertical side members.

The first sizeable tunnel in soft ground is recorded by Sandström (1963) as the Tronquoy tunnel on the St Quentin canal in France in 1803, where the method of construction, based on the use of successive headings to construct sections of the arch starting from the footing, was a forerunner to the German system described above.

1.4 New methods, tools and techniques

Rock tunnelling was revolutionised by the first use of explosives for mining in Germany in the seventeenth century (Sandström 1963) and by the use of compressed air for pneumatic drills in 1861. Previously, the methods had been laborious and slow, scarcely improving on principles used by the Egyptians and later the Romans who understood how to use plugs and feathers, also the expansive effects of wetted wooden plugs driven into drill-holes.

The Fréjus Tunnel (also known in Britain as the Mont Cenis Tunnel), the first Alpine tunnel, constructed between 1857 and 1870, first used compressed-air drills – drilling prior to this time a laborious operation by hand-operated drills (using 'jumpers' and hammers). A steam-driven rock drill, invented by a British engineer, T. Bartlett, was successfully adapted for compressed air in 1855 but was not developed commercially. Several features of other proposals, including that of Joseph Fowle of the United States, were incorporated in the rotary-impact drills with automatic feed used by G. Sommeiller for the formidable task of driving the 12 224 metres long Fréjus Tunnel. The first compressor installation of 1861 used falling water with entrained air, separated under a pressure of 700 kPa, but excessive dynamic stresses caused unreliable operation and a water-wheel was substituted to power the compressors. The first drilling carriage ('jumbo') was also used for this tunnel, mounting four to nine drills. Fréjus is also notable for the consideration given by Sommeiller for the accommodation, medical and school services provided for the work force.

Soft ground tunnelling was liberated from its constricted use by the invention of the shield by M.I. (Sir Isambard) Brunel, who, at the same time, introduced other expedients which later became widely adopted (spiles [renamed by some in the 1980s as soil-nailing], methods of face support), demonstrating a clear, although yet largely qualitative, understanding of the interaction between the ground and its support, also the means for controlling entry of water. His contribution to tunnelling was so remarkable that it merits more than a brief mention in this chapter.

Brunel's Patent Application of 1818 (Brunel 1818) depicts two types of circular shield, one advanced by propelling rams bearing against a reaction frame, itself blocked against the most recently completed length of the tunnel lining. The other type of shield comprises a central rotating drum while the lining is advanced as a series of segments built in a helix (Muir Wood 1994a). Brunel was thinking ahead of the current practicability of incorporating hydraulic machinery into a moving shield.

The first shield-driven tunnel, the Thames Tunnel, initially designed predominantly for horse-drawn traffic serving the docks under construction on the south bank of the River Thames, was however rectangular. Muir Wood (1994a) describes the shield in considerable detail. Essentially, it comprised 11 (later 12) vertical cast-iron frames each containing three cells, each about 2 m high × 0.8 m wide, one above the other. At the head, and to each side of the shield, a series of 'staves' with sliding and pivoted support from the frames, had cutting edges at their leading end, and were extended by wrought-iron skirts at the rear to permit the brickwork lining to be built within their protection. The frames and the staves could be independently propelled forward by screw-jacks bearing against the last completed section of brickwork. Initially the section of the excavated tunnel was 10.83 m (35 ft 6 in) wide by 6.25 m (20 ft 6 in) high, increased to 11.44 m (37 ft 6 in) wide by 6.71 m (22 ft) high for the second part of the tunnel.

The face was supported by elm poling boards held against the face by screw-jacks. Excavation was achieved by the removal of one or more poling boards in each cell. Ultimately successful, after an ordeal which would have daunted a lesser engineer – arising mainly from unexpected features of the ground, from consequent irruptions by the river and the great difficulty in the essential coordination of all the activities concerned with the advance of the shield in appalling conditions and poor light – tunnelling started in 1825, and was completed in 1841. By this time the nature of the requirement had changed so that the tunnel was initially opened for pedestrians in 1843 and for the East London Railway (now part of London Underground) in 1869.

The quality of bricks and the design of the Roman cement mortar used by Brunel were designed to limit to a minimum inflow of water. In order to drain any water seeping through the brickwork, Brunel provided a system of circumferential slots cut or otherwise formed at the intrados of the brickwork, at intervals of about 0.2 m, with the internal lining provided by tiles and a plaster rendering, illustrated by Roach (1998) who describes work undertaken to provide an internal concrete lining in 1996 as a replacement to the original internal lining. The main principle of conservation was assured at this time by causing minimal change to the stress regime of the original brickwork, predominantly caused by water loading, by means of adequate and readily maintainable drainage between brickwork and reinforced concrete lining (which may well itself require repair during the future period of expectation of life of Brunel's tunnel).

The first circular shield, on the basis of a patent by P.W. Barlow, was used in 1869 for the Tower Hill subway tunnel beneath the River Thames, for which J.H. Greathead, assistant to Barlow, designed a simpler version propelled by screw-jacks. The tunnel was first to be lined in cast-iron segments adopting the illustration of such a lining in Brunel's 1818 patent. The tunnel, 1340 ft (408 m) long, had an internal diameter of 2 m with a lining flange depth of 75 mm. The lining was erected within the skirt of the shield and grout was injected to fill the annulus behind the lining by syringe. The first use of hydraulic jacks was for the shield built by A.E. Beach for a tunnel in 1869 under Broadway, NY, otherwise based on Barlow's patent.

Greathead continued to develop many ideas for exploiting the potential of the shield, including the closed-face shield with the ground broken up by jets and the spoil removed as a slurry, the forerunner of the slurry shield, first used at New Cross in London in 1971. He also proposed the trapped shield, for use in unstable ground enabling, in the event of a face collapse, stability to be restored through pressure balancing on a horizontal surface between the two partitions forming the trap. Greathead also designed the grout-pan for continuous mixing and injection of grout, a device in use to the present day. The City and South London Railway of 1886 (from City to Stockwell – now part of the Northern Line of London's Underground) provided the first opportunity for Greathead (Greathead 1896) to demonstrate the potential of improved shield tunnelling. These twin 10 ft 2 in (3.1 m) diameter tunnels, each 3.5 miles (5.6 km) long, used electric locomotives for traction and compressed air where required by the ground. The use of compressed air for balancing water pressure, including the use of horizontal air locks had been proposed by Lord Cochrane's Patent 6015 of 1830. Compressed air had been widely used for diving bells and for caissons, e.g. for bridge piers, where the balance was achieved through a horizontal water surface, a tunnel presenting the more complex problem of a varying pressure over the vertical height of the face. A.M. Hersent successfully constructed a small cast-iron heading, 5 ft (1.5 m) high by 4 ft (1.2 m) wide, in compressed air in Antwerp in 1879.

Compressed air was used for the Hudson River Tunnel by D.C. Haslin in 1879. This pair of railway tunnels was to be lined in cast-iron segments with internal brickwork to give a finished diameter of 18 ft (5.5 m). The tunnel was entirely in silt which stood at a stable slope of about 45 degrees in the air pressure, cut in benches, without other face support or protection. The first disaster occurred when attempting to link the two tunnels by a heading close to the main working shaft. The work was recovered, an improved system devised with a short length of pilot tunnel, but the whole abandoned in 1882 when the money ran out. The tunnel was subsequently advanced with the use of a shield with hydraulic segment erector, by Sir William Arroll advised by Sir Benjamin Baker. Financial problems, however, prevented completion until 1899 (Jacobs 1910).

Greathead shields were also used for two subsequent road tunnels beneath the River Thames, at Blackwall (1892–97) (Hay and Fitzmaurice 1897) and Rotherhithe (1904–8) (Tabor 1908).

The technique of tunnelling in open ground continued to require a closely timbered heading ahead of the shield, no great advance on the technique used by Brunel. In 1896 Sir Harley Dalrymple-Hay developed the hooded shield with clay-pocketing, the packing of a puddled clay into a series of adjacent holes forming an annulus ahead of the cutting edge of the shield (Dalrymple-Hay 1899). Struts supporting the face timbering allowed the shield to be advanced into the clay annulus while face support remained undisturbed. A similar technique, using steel shutters supported by screw jacks (once again recalling Brunel), was used for the Blackwall Tunnel, for which air-locks built into the shield, never in fact used, would have allowed differential pressures for the upper and lower sections of the face, evidently requiring a substantial seal at the intermediate working platform.

For the main drainage system for London of Bazalgette (Bazalgette 1865) sub-drains were widely used to draw down ground-water which would otherwise have caused slumping of the face of a tunnel in water-bearing ground. A comparable technique was used for the Mersey rail tunnel (1881–6), the low-level drainage tunnel in sandstone excavated by a machine designed by Beaumont.

The developing interest in the Channel Tunnel inspired many proposals for mechanical excavation (Lemoine 1991). The first trials of a machine of J.D. Brunton, using disc-cutters in 1882, achieved 12–17 m/day but its design presented unsolved maintenance problems. Colonel Frederick Beaumont invented his machine, a rotary machine with picks, in 1875. The design was improved by Colonel Thomas English and tested for the Channel Tunnel in 1880. By Spring 1882 this 2.1 m diameter machine had bored a tunnel 800 m long at Abbotscliff and a tunnel 1100 m long under the sea from a shaft at Shakespeare colliery (later to form part of the main working site for the Channel Tunnel). An improved and more powerful machine, built in 1882 by Batignolles, excavated about 1700 m of the total length of 1840 m of the French tunnel at Sangatte, achieving a maximum rate of advance of 24.8 m/day. In 1883 T.R. Crampton proposed a machine, hydraulically driven by water from the sea, which would also be used to evacuate spoil as a slurry. A heading 150 m long was driven at Folkestone Warren in 1922 by a machine designed by Douglas Whitaker, lightly built by present standards. The further development of tunnelling machines had then to await the 1950s.

Copperthwaite (1906) and Simms (1944) provide contemporary accounts of practical developments seen from a British viewpoint. For a brief account of the major developments in Britain since 1875, see Muir Wood (1976).

The first of the generation of mechanical shields used the Price Digger of 1897, with chisel picks attached to arms rotated by electric motor, which was followed by numerous comparable designs. Table 1.1 indicates the approximate rates of progress achieved over the years for driving running tunnels in London. Rotary shields in clay with expanded linings have permitted rates of advance for sewer and water-supply tunnels in London clay and Gault clay well in excess of 1000 m/month. For tunnels of relatively short length, the use of road-headers within simple shields have achieved rates in excess of 350 m/month (Jobling and Lyons 1976).

The process of placing concrete *in situ* was incompatible with timber supports. In consequence, the first uses of concrete were for tunnels in good rock and it was only with the introduction of steel supports that the use of concrete became the norm for a tunnel lining material. An early use of concrete in tunnel lining was for the East Boston Tunnel of 1892,

Table 1.1 Approximate rates of advance of running
tunnels for London's underground railways

Year	Rate of advance (m/week)
1876	3.5
1886	12
1939	50
1960	90

where the tunnel was lined in stages, a crown shield supported on the concrete side-walls. In the early years of the twentieth century, concrete began to be used more generally for tunnel lining, the concrete placed behind timber formwork supported on steel arches, taking advantage of *in situ* concrete to mould itself to the shape of the rock extrados. Steel arches had largely superseded the use of timber in rock tunnels and, later, rock bolts were introduced from mining.

Cast iron for internally flanged linings, particularly of circular tunnels in soft ground, often constructed in shield, became widely used, using the material in such a manner to take advantage of its strength in compression and weakness in tension. As cast-iron linings increased in cost, so were alternative materials sought. Linings of rings of bolted, flanged, reinforced concrete segments were introduced in 1936 (Groves 1943), used to the present day for hand-constructed shafts and tunnels and for primary linings. The next major development had to await expanded linings, described in Section 1.5 below.

1.5 Towards the present day

Until the 1950s tunnelling continued to be seen as largely a traditional craft-based operation, undertaken by skilled miners and timbermen largely educated from experience. Design methods, developed in the nineteenth century, were being extended but not essentially modified by Kommerell (Figure 1.2), Terzaghi and others. The time was ripe for the application of the rapid developments in the understanding of the behaviour of soils and later, and more partially, of the behaviour of rocks. It had long been understood that the ground, if allowed to deform slightly, was capable of contributing to its own support. With the general replacement of timber by steel supports, it was now becoming possible to quantify this effect and to design the construction and timing of support correspondingly. Apart from the positive influence from the advances of the supporting technologies, demand for tunnels suddenly increased as the world began to recover from the material shortages and economic constraints following

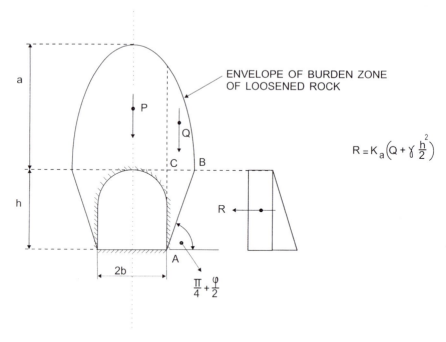

Figure 1.2 Rock loading on a tunnel (after Kommerell 1940).

the Second World War, for road/rail (including Metro) transport, tele-communications, water and sewage conveyance, oil and gas storage, hydro-electric and defence purposes, recognising the increased penetrating power of missiles.

Ground exploration techniques (see Chapter 4), partly derived from oil exploration, developing from the 1930s, allowed recovery of (relatively) complete and undisturbed samples of soil and cores of rock. Soil mechanics as a recognised technological specialism began to permit the principal properties of the ground in terms of strength and deformability to be applied to the design of tunnel excavation and support. Some years later – the delay a consequence of rock types spanning a wide range of varieties of behaviour as a continuum/discontinuum – rock mechanics was applied to the design of arch support (Terzaghi 1961) and subsequently by Lang (1961), Rabcewicz (1964) and many others to the use of rock bolts, sprayed concrete and allied means of initial support. In view of the complexities of rock as a material, advances were largely dependent on bold experiment, often at full scale, followed by a degree of conceptual-isation and analysis, supplemented by laboratory experiment. This is a far more typical manner of advance in civil engineering than the traditional notion that the order runs: theory; experiment; development, application;

trial and demonstration, as in some branches of science. Civil engineering is in this respect akin to medicine. Jaeger (1955) summarises the experiences in pressure tunnels, including successes and failures.

A fundamental development for tunnels in soft ground was the appreciation that generally, and this could be readily if only approximately estimated, the strength of the ground was adequate, with some degree of radial support from the lining, to resist (elliptical) deformation of a circular tunnel, as a result of differences between vertical and horizontal ground loading. From this feature follows the conclusion that the ideal lining is capable of adequate strength to develop circumferential force while having minimum stiffness in bending in the plane of the ring (Muir Wood 1975a), characteristics which could be achieved, for example, by means of the use of segments with shaped, initially convex/convex longitudinal joints (see Figure 5.14), care being taken in the design to avoid the risk of high local bursting stresses. Jacked linings or linings with soft packings in the joints could also, subject to the practicalities of construction, serve to match the convergence of the ground to reduce the load in the ring. While initially such concepts were applied to linearly elastic ground and linings, subsequent availability of computer power with the ability to introduce full constitutive relationships of the ground, based on the concepts of critical state, have permitted much refinement of these early approaches. Current applications are described in Chapter 5.

Traditionally, segmental linings had often been used in conjunction with a tunnelling shield, the segments being erected and bolted together to form a ring within the protection of the skirt of the shield, the resulting annular space being filled with cement (earlier, lime) grout. For tunnels in clay capable of standing unsupported briefly over the width of a ring, the use of linings built directly against the ground simplified the process and allowed greater rates of progress to be achieved. Where the ground could be cut to a smooth cylindrical surface by the shield, the lining could be erected in rings and expanded against the ground, thereby minimising inward ground movement. For this purpose, the most satisfactory method for small diameter tunnels entailed the use of tapered segments so that the ring was expanded by driving such segments into the ring by the use of the propelling jacks of the shield. The Don-Seg lining (Figure 1.3) was first used for the Lee Valley–Hampton water tunnels following their successful use in an experimental tunnel in 1951 (Scott 1952, Cuthbert and Wood 1962). The improved rates of construction achieved by the several innovations are illustrated by Table 1.1. Donovan (1969) describes developments in segmental linings from British experience.

For large diameter tunnels, where a high rate of progress was not a primary aim, the lining was expanded by means of jacks inserted in the ring between segments (Figure 1.4), with the jacks replaced by dry concrete rammed in place in a sequential manner, a practice with the potential of

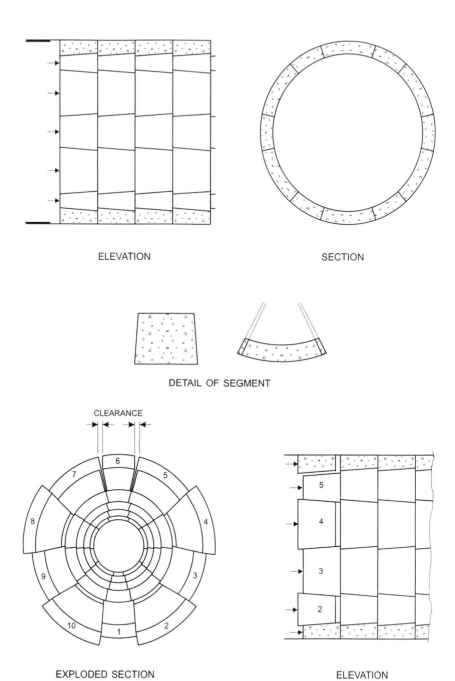

ELEVATION

SECTION

DETAIL OF SEGMENT

CLEARANCE

EXPLODED SECTION

ELEVATION

Figure 1.3 Don-Seg expanded concrete lining.

preloading the lining sufficiently to reduce settlement. The Potters Bar railway tunnels provided in 1955 (Terris and Morgan 1961) an opportunity to design an expanded lining for a tunnel of 26 ft 6 in (8.1 m) internal diameter. The tunnel was to be used by steam locomotives so a sulphate resisting cement, Sealithor, was adopted for the precast segments. Provision needed to be made for frequent refuges and for the throats of smoke shafts, also for uncertain and weathered ground at each portal of the three tunnels. A lining of constant thickness was used, adequate to provide for each of these special circumstances (Figure 1.4). Each ring comprised a massive invert segment, with 19 'morticed and tenoned' voussoirs to complete the ring, 18 in (0.45 m) wide, 27 in (90.68 m) thick. At each side, at axis level, pockets were provided for hydraulic jacks, used to expand the ring immediately after erection, designed to occur as soon as the shield completed a forward shove. At refuges, openings were formed by this procedure, an adaptation of tradition:

1. above and below the refuge, voussoirs cast with semi-cylindrical radial pockets at adjacent faces were built into the lining such that 150 mm cylindrical holes were provided through the full thickness of the lining;
2. earth-dry fine concrete was compacted into the cylindrical cavities to provide shear keys between rings;
3. temporary segments were withdrawn in the position of the refuge which was completed by further hand excavation and provision of an arched brickwork rear wall.

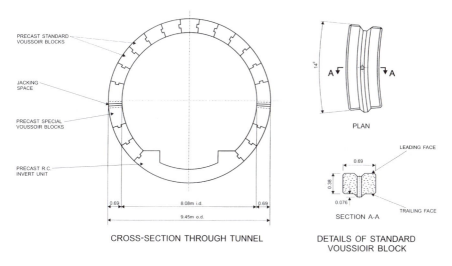

Figure 1.4 Potters Bar tunnel lining.

For the smoke shaft throats, each the width of four rings, i.e. 6 ft (1.8 m), a more complex procedure was required, adopting (1) and (2) above supplemented by the threading of longitudinal Lee–McCall prestressing bars to avoid risk of spreading of the 'lintel' and 'sill' above and beneath the opening.

Several means of expanding the ring were tested in an experimental tunnel for the Victoria Line of London's Underground in 1960–61 (Dunton *et al.* 1965), including the use of internal jacks with the insertion of folding wedges and the use of a wedge segment in the crown. For the Victoria Line (Follenfant *et al.* 1969, Clark *et al.* 1969), jacks were used at 'knee joints' with precast blocks inserted in the space.

London Heathrow Airport's Cargo Tunnel (Muir Wood and Gibb 1971), a two-lane road tunnel, passes beneath Runways 5 and 6. Great economy could nevertheless be achieved by designing a tunnel for low cover ($C/D < 0.78$). An expanded lining was adopted for this 33 ft 9 in (10.3 m) internal diameter tunnel, illustrated by Figure 5.14, using segments 12 in (300 mm) thick and 24 in (600 mm) wide, the ring being prestressed by jacks at axis level. As for the Potters Bar tunnels, after full expansion dry concrete was packed to each side of the jacking space, the jack subsequently removed and the concrete packing completed.

The Potters Bar lining had been deliberately made stiff, i.e. with bearing across virtually the full width of each longitudinal joint. The benefits of articulation were achieved elsewhere by shaping the longitudinal joint, generally as convex–convex arcs to permit rolling without the risk of one segment 'hanging up' on another away from the centroid of the section. High precision in casting ensured the benefits of the design without risk of local excessive stress, tolerances of 0.1–0.2 mm being demanded and achieved, together with complex angular tolerances, using specially designed jigs. By the time of the Channel Tunnel, high precision 3-D metrometers were available, undertaking the computation to establish that dimensional limits were being achieved (Eves and Curtis 1992).

For tunnel linings through ground imposing uneven loading, secondary bending stresses need to be accepted. Grey cast iron, traditionally used for tunnel segments, has limited tensile strength. In consequence, for such situations, spheroidal graphite (s.g.) iron, with tensile strength approaching that of steel, has been widely used as a bolted segmental lining, for several Metro tunnels and for parts of the Channel Tunnel (Eves and Curtis 1992). Steel segmental linings have also been used for special circumstances of abnormal loading (Craig and Muir Wood 1978) where casting procedures would be uneconomic. The evolution of the design of tunnel junctions has developed from cast-iron frameworks (the so-called 'picture-frame' opening) to many forms of the use of steel and concrete for the purpose.

For many of the early examples of expanded linings, the segments were designed of such dimensions as to permit handling, placing and stressing without the need for reinforcement, for reasons of economy, to assist in achieving dense high quality concrete and to avoid long-term problems of corrosion. As such linings have encouraged high-speed tunnelling, so has the use of wider rings (1–1.25 m) developed, with the consequent need for reinforcement and for additional protection, by particularly dense concrete, careful control of depth of cover to reinforcement, the use of cathodic protection or epoxy coating of reinforcement, particularly where the lining is exposed to saline or polluted water.

Rock tunnelling evolved, following the Second World War, from a traditional craft using familiar materials in a familiar manner, into a technology-based art. Ground description was developing towards quantification (Talobre 1976) permitting emerging attempts at analysis, an increasing choice of means of support and an appreciation of the influence on the behaviour of the ground of the procedures in working. The sheer complexity and variability of rock impeded a rational approach; on the other hand, it was becoming evident that any rationale would need to be expressed in simple terms or variability would require continuous reassessment. Here it is worth reflecting that, in many circumstances, a length of tunnel benefits from the contribution of the third dimension since a local failure mode must cause work to be done, as 'membrane failure' of the tunnel lining and by contributions of the ground, a mechanism not reflected in a 2-D analysis.

Several engineers through the nineteenth century, such as Ržiha (Kastner 1971) were pointing to the benefit of early support for jointed rock to anticipate high loading caused by rock loosening as a consequence of excavation. Only later, however, as Kovári (1999) describes, Wiesmann in 1912 established, in a qualitative manner from experience of the Simplon Tunnel, the beneficial consequences of allowing deformation of squeezing rock. The corollary was therefore to relate the properties of the support to the stress/strain behaviour of the rock around the advancing tunnel in order to provide an economic and stable solution. This concept is now accepted as fundamental to tunnel design, with great consequent economies in tunnel support. There is also a hidden benefit in that the risk of sudden brittle failure is thereby reduced. The approach towards this end has seen a considerable variety. From one viewpoint, the engineering geologist (Talobre 1976) has seen an understanding of the geological structure as the key to success. From another viewpoint, the theoreticians (e.g. Kastner 1971, Ladanyi 1974) have endeavoured to express the loadings as closed-form solutions as a basis for the geometrical design of the tunnel, providing that reliable ranges of values may be given to the variables. In between, observation in tunnels using innovative forms of support, assisted by understanding the geology and using analytical techniques in simplified form, allow pragmatic rules to be developed for comparable circumstances. The authors of Rock Mass Classification systems have endeavoured to provide simplified keys to the generalisation of such observations. Present and possible future developments of convergence between these approaches is discussed in Chapter 5.

For jointed rock, Engesser in 1882 (Széchy 1970), followed in 1912 by Kommerell (1940) had provided a simplified rational basis for estimating the rock load to be supported by an arched tunnel (Figure 1.2). Terzaghi (1961) developed this approach for steel arch supports with suggested values for the rock strength parameters. The main problem in design of steel arches was that of discontinuity of contact pressure between rock and arch, traditionally achieved by timber packing, with consequent indeterminacy in calculating collapse load for failure by torsion and lateral buckling. The expedient of a porous bag filled with weak mortar between rock and flange of arch (Muir Wood 1987) much reduced the uncertainty and increased the load-carrying capacity (Figure 5.13).

More generally, appreciation of the merits of light support with properties of deformability to suit the ground have long been a feature of the approach of many engineers, by some as a reaction to the experiences of the massive masonry linings, rigid and slow in construction, of the early trans-Alpine tunnels, where the permanent structure might have occupied as much as 40% of the excavated profile (Müller 1978, Fechtig and Kovári 1996). The common objective has been to enable the ground to be self-supporting around the periphery of the tunnel to the greatest degree. Mining practice provided much of the incentive for liberation from tunnelling tradition, mining being relatively unconcerned about questions of loss of ground, hence capable of exploiting the natural properties of coherent ground, making use of rock-bolts, sprayed concrete and road-headers for excavation. The first two provided components for rapid, light, yielding support, the latter a form of mechanised excavation adaptable to wide variation in geometry. Dubious claims are made that as early as 1944, Rabcewicz (Rabcewicz 1944) foresaw the benefits of tunnelling practice based on such features. In fact, there is a long history of the perception, application only becoming practical with the later development of tunnelling techniques, in response to demand.

Here again the major steps forward in tunnelling occurred in the 1950s with the principles coupled with growing understanding of the behaviour of jointed rock, the availability of new techniques in rock-bolting, the use of rock anchors and the application of sprayed concrete (Shotcrete being a trade name) and, most importantly, the opportunity to test out new ideas with growing confidence in practical tunnelling. Two noteworthy early developments were in Switzerland, 1951–55, using Shotcrete for the Lodano–Losagno Tunnel for the Maggia hydroelectric project and the Snowy Mountains Hydroelectric Project in New South Wales, Australia from 1952 (Lang 1961, Andrews *et al.* 1964, Dann *et al.* 1964, Moye 1965). Many others in many other countries have grasped the essence of the principles and applied it with success.

Rabcewicz coined the title 'New Austrian Tunnelling Method' (NATM) in 1962 (Rabcewicz 1964) with an explanation:

'... the author carried out during the (1939–45) war a new method called the "auxiliary arch" which consisted of applying a relatively thin concrete lining [a conventional in-situ lining of the time would be an unlikely candidate for the "observational method" then described] to the rock face as soon as possible, closed by an invert and intended to yield to the action of the protective zone [the name given by Rabcewicz to the rock adjacent to the tunnel]. Deformation of the auxiliary arch was measured continuously as a function of time. As soon as the observation showed a stabilising trend of the time/deformation curve, another lining called the "inside lining" was constructed inside. The method can be considered as a real predecessor of NATM as it comprises all the integral factors with the exception of modern means of surface stabilisation.'

If the approach (it is not a method) had retained comparable simplicity of principle, the term NATM would be more generally accepted at the present day as defining a valuable step in furnishing the tunnel engineer's tool-kit (Müller 1978, Müller and Fecker 1978). But the succeeding generation of Austrian tunnelling engineering could not forbear to invest the principles in an impenetrable shroud of complexity – and downright nonsense – possibly with an aim of combining market penetration with apparent academic respectability. Examples include the extraordinary variety of definitions of 'ground rings' (the 'protective zone' of Rabcewicz), i.e. the zones of rock naturally stressed by excavating the tunnel, which girdle – or in some exigesises appear only partially to girdle – the tunnel. In reality, any opening in the ground causes a degree of ground support to be conferred by circumferential stress of the ground. The official definition of NATM, issued by the Austrian Society of Engineers and Architects reads:

The New Austrian Tunnelling Method constitutes a method where the surrounding rock or soil formations of a tunnel are integrated into an overall ring-like support structure. Thus the formations will themselves be part of this supporting structure.

This, as Kovári (1993) explains, is a universal consequence of tunnelling.

Another curiosity concerns the notion of a minimum point of the Fenner–Pacher ground response curve relating convergence to the degree of support required (see, for example, Brown 1981) – the forerunner of the confinement-convergence curves. (Rabcewicz 1969, Rabcewicz and Golser 1973) – which has become established in NATM lore, contrary to common-sense. The correct timing of the provision of support is said to be a fundamental principle of NATM but without a minimum point for support need this criterion does not exist. Much more important becomes the need to secure the ground before the rate of convergence becomes excessive, a feature omitted from the Fenner–Pacher concept (apart from an extraordinary multi-dimensional version proposed by Sauer (1986) which appears to be devoid of theoretical basis). The consequence has been, quite predictably, much confusion as to a definition of NATM as a result of banality of the philosophical attributes which are claimed for it (Kovári 1993). A rational approach to systems of Informal Support is discussed further in Chapter 5.

Practically, for the 'brave' new alpine tunnels, light support capable of tolerating a high degree of inward convergence of highly stressed rock to establish equilibrium, has achieved great economies by comparison with traditional methods. This was achieved for the Arlberg and Tauern tunnels (John 1980) by the use of yielding Toussaint-Heinzmann arches (see Section 5.2.1) since the degree of convergence would be excessive for a load-bearing Shotcrete lining. Additionally, in weaker ground, the techniques of building a side- or crown-heading directly into the full tunnelled section draws upon earlier traditional practices with steel arch supports, in other than squeezing ground (Kovári 1998). For a rock tunnel under high ground loading, the principal objective will be to establish stability while limiting the extent of the loading to be transferred to the support or to any subsequent lining. For an urban tunnel in soft ground, the principal objective is different, namely to provide support so expeditiously as to secure the ground and to limit ground movements which may otherwise be expensive to compensate or cause damage to other structures and services. Reduced loading on the support in soft ground is natural and adventitious as a result of the deformation of the ground ahead of the installation of support (Panet and Guenot 1982).

It is a universal principle of good tunnelling to adapt the support to the expected behaviour of the ground. On the one hand, NATM has been used to cover a variety of different approaches to tunnelling and, on the other, many engineers approach tunnel design in a rational manner which does not wish to wear impenetrable philosophical baggage. In this book, the term 'Informal Support based on Monitoring' is the general description, contracted to Informal Support, emphasising that this title includes all forms of tunnelling which entail the provision of support intermediate between excavation and a final formal lining (which might itself be in sprayed concrete). Support may include the use of bolts, spiles, dowels, arches and sprayed concrete. Here 'Informal' implies a variability and capability to respond to needs as these may vary between tunnels and between zones of a single tunnel. For the practice of tunnelling in yielding ground with light support, the description of Sprayed Concrete Lining (SCL) is used, following the lead of others (Institution of Civil Engineers 1997), with these features:

- initial support designed on the basis of the composite behaviour of the ground and the support;
- a reliance on observation (including the use of appropriate instruments) as a basis for establishing adequacy of support (or, for circumstances where time allows, the need for supplementary support);
- unified management of the design/construction stages of excavation, initial support, monitoring and final lining.

There are engineers at the present day who understand SCL to exclude reliance solely on rock-bolting, which seems reasonable in a logical sense. Informal Support is a description of universal application, however. Early examples of SCL were based on simple analysis of a tunnel in ground of uniform initial stress, considering the ground as a linear elastic material up to yield and as a material of uniform plastic strength thereafter, as described by Kastner (1971) and others. Subsequently, the effects of anisotropy and of the strength reducing to a residual value have been considered. For yielding rock, the approach remains largely empirical, comparing one rock against another, initially by 'scoping' and then by considering experiences with similar rocks elsewhere before calibrating behaviour in direct trials as tunnelling begins (Kidd 1976). Analysis may be kept simple so long as the rock is of one type and there are no hidden hazards of the site, such as confined aquifers, which might prevent timely undertaking of remedial work.

There have been many developments in different forms of rock-bolt, spiles and dowels to suit different circumstances, the current choice being described in Chapter 6. Problems of 'shadowing' of sprayed concrete by traditional steel arches encouraged increasing use of lattice arches formed of bar and mesh. The techniques of sprayed concrete have been developed to include robotic application and the inclusion of steel fibre reinforcement, while the debate of the merits of dry and wet application continues, as described in Chapter 6.

An alternative scheme of providing immediate ground support prior to excavation, developed in France, *sciage*, creates a slot in the ground around the projected extrados of the tunnel, the slot being filled with sprayed concrete which develops enough strength to support the periphery of the ground prior to the main excavation (Bougard 1988). Lunardi *et al.* (1997) describe recent use of the method for a 21.5 m span of a Metro station roof.

The earliest rotary tunnelling machines date from the 1880s, including the machine attributed to Beaumont and used for the lengths of pilot tunnel advanced for the Channel Tunnel from the coasts of France and England. Mechanical shields for soft ground were largely developed for working in the London clay (Stack 1995) with the consequent improvements in performance illustrated by Table 1.1. Further developments were slow until the spur of demand in the 1950s, when large rotary rock boring machines and 'closed face' mechanical shields started to be developed, the latter relying on the use of a thixotropic medium, initially bentonite, in the front of the machine to stabilise the face of the ground (Bartlett *et al.* 1973). The simple differentiation between mechanical shields for soft ground and rock tunnelling machines for rock began to break down, the term Tunnel Boring Machine (TBM) becoming a generic descriptive term for all species of machine with full-face rotary cutter, the sub-species being represented as:

- open TBM for rock tunnels, i.e. having no cylindrical shield to support the periphery of the ground;
- shielded TBM for weak rock (or jointed rock with low cover) and soil, the shield providing an intermediate ground support between the face and the support provided behind the TBM;
- TBM with closed face, supporting the ground by means of slurry filling in the front cell, created from the excavated ground mixed with bentonite, polymer or a foaming compound;
- earth pressure balancing machine (EPBM), where support is by excavated spoil contained under pressure in the front cell of the machine.

There are a number of different forms of the last two types with intermediate versions. Spoil of the slurry type machine is usually removed by pumping from the face cell, whose pressure is stabilised through an upper cell forming a compressed air vessel. Spoil from the EPBM is removed by inclined Archimedean screw, often in two sections to give greater control by differential rates of rotation. See also Sections 5.1.5 and 6.3.

Where a segmental lining is constructed at the rear of the TBM, it is often the practice to articulate the machine, with intermediate jacks between the forward and rear sections, so that the front section may continue to advance as the lining is being erected. This practice also helps the ability to steer the machine.

Pipe-jacking (also known as thrust-boring) is first recorded in the USA in 1892, originally providing an iron sleeve for a service pipe, later using steel or concrete pipes for the purpose, under a road or railway line. Subsequent developments world-wide, but principally in the USA, Europe and Japan, have adapted the method for long lengths of pipe and tunnel. While pipe-jacking essentially entails application of a jacking force from the point of departure of the tunnel, several features have contributed to the extension of the capability and its applications:

1. the design of abutting joints between pipe elements, the earlier rebated joints being generally superseded by butt-ended joints with dowels or sleeves to maintain alignment, while controlling loading between pipes and excluding the ground;
2. the use of intermediate jacking stations (IJS), thus limiting the jacking load to advance the tunnel, allowing lengths of jacked tunnel or pipe in excess of 1000 m;
3. the use of external lubricants to reduce frictional forces opposing forward movement;
4. the use of mechanical excavation, ranging from shields with excavators to closed forms of TBM, at the face of the tunnel;
5. the ability to introduce curves into a pipe-jacked tunnel.

Since 1970, miniaturisation, mostly undertaken in Japan, has developed a family of devices for cutting and reaming the ground, largely superseding the earlier systems which depended on jacking a blind pipe. Recent developments are discussed in Chapter 6.

Drill-and-blast techniques have advanced through the use of multiple, servocontrolled hydraulic drills, the process being increasingly computerised overall, the development of explosives and delays to suit the requirements for the main rock-breaking and for the smooth-blasting around the periphery, also the adoption of non-electrical forms of firing (c.f. the fuses of the early uses of gunpowder).

Road-headers and boom-cutters of several types, powers and sizes have been developed from their widespread usage in mining. The ability to excavate strong rock and to limit wear of the picks depends as much on the overall stiffness of the machine, (i.e. the combination of mass, limited mechanical clearances between elements of the system, sturdiness of design and support of the machine to minimise 'chatter') as on its power.

Transport of spoil within the tunnel may determine the speed of advance of the tunnel. Long extensible conveyor belts have been used for this purpose where space and geometry allow, but more often trains of wagons, either overturned into hoppers at working shafts or equipped with floors which open at the transfer station. For confined spaces in small diameter tunnels, a long rail-mounted wagon with the floor formed of armoured conveyor (Salzgitter car) has been widely used. Where the spoil is fine or where it can readily be treated to be pumpable, the spoil has been pumped from the TBM to the point of disposal, this particularly for closed-face machines, requiring recirculation of the stabilising agent. For the French section of the Channel Tunnel, on the other hand, spoil was transported by rail-mounted wagons to the principal shaft where, after pulverising in a mill and mixing with water, the spoil was pumped to disposal in a surface pond behind an earth dam (Barthes *et al.* 1994).

Another area of major development has been that of treating the ground in order to establish stability or the control of inflow of ground-water ahead of the advancing tunnel. Such methods are here referred to collectively as 'special expedients'.

Ground freezing by the use of brine in circulation pipes was first used for shaft-sinking for mining through water-bearing zones of the ground in South Wales in 1862 (Glossop 1968). More recently, the technique has been adapted for tunnelling, using a pattern of freezing around a cylinder ahead of the tunnel together with a plug of frozen ground at its extremity. The tunnel is then advanced in stages to allow the next cycle of freezing to start, a relatively slow process. An alternative method of freezing has used liquid nitrogen as the circulating fluid, requiring less mobilisation of plant and providing a more rapid freezing process, hence allowing higher overall rates of advance and better able to cope with the heat transport problems of moving ground-water. The method has usually used vertical or near-vertical freeze pipes from the ground surface to freeze a block of ground through which the tunnel is to pass, the excavation for the tunnel cutting through any, previously decommissioned, freeze pipes which are intersected. The process advances with the tunnel until the waterbearing area is passed.

Treatment of the ground by injection may be required to increase the strength of soils, to increase the stiffness of jointed rock (Jaeger 1955) or to make tight those joints in rock otherwise contributing to instability. Treatment is more often used to reduce the entry of water from open-textured soils or open-jointed rock by reducing its mass permeability. Traditionally, cement grout has been the medium, limited in its use to soils of coarse sand size and above. The range of available materials has increased over recent years to include bentonite-cement, fine micro-cement, silicate grouts and a wide range of resins with controllable setting times. Essentially, the finer the soil (or the fissures in rock), the more elaborate the overall grouting process and the more expensive the types of grout needed to achieve results by penetrating the finest fissures.

Glossop (1968) recalls that grouting of a primitive nature has been practised over very many years, that Bérigny introduced grouting with clay and hydraulic lime in 1802 and that Hawksley used cement grouting for dam foundations from 1867.

The cement grouts set as a result of wetting, the other grouts requiring setting agents. The traditional silicate grouts, whose setting characteristics depend on pressure, concentration and temperature, may tend to break down (syneresis) prematurely and should only be relied upon over short periods. Some of the resins (and the earlier cyanide grouts no longer in general use) are highly toxic and demand great care in their use. Harding

(1946) describes early uses in Britain of the Joosten process of grouting, based on silicates with an organic reagent, resulting in a gel of higher strength than the traditional silicate grout.

For tunnelling, while ground treatments have been applied from the surface for shafts and for shallow tunnels, the usual expedient, where long continuous lengths of tunnel call for treatment, is to form overlapping cones of treated ground from a pattern of drill holes diverging at an acute angle to the axis of the tunnel, in an aureole ahead of the advancing tunnel or from a pilot tunnel (Kell and Ridley 1966). For drill-and-blast and hand-tunnelling techniques, the holes for the purpose are drilled at the face. Where a TBM is used the aureole needs to be drilled from the rear of the machine, requiring appropriately increased length of overlap between drill holes.

The use of *tubes-à-manchette* for grouting has become widespread, allowing by means of a grouting tube and an external casing (Figure 1.5), the injection of grout of known volume into a selected zone of the ground.

As recalled above, the use of low-pressure compressed air in tunnelling dates from 1879 for the Hudson River Tunnel, the pressure of the air

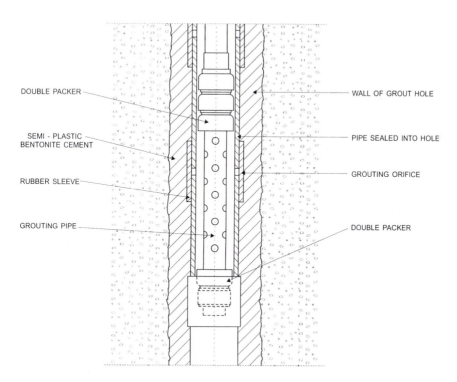

Figure 1.5 Tube-à-manchette.

serving to reduce the entry of water and occasionally also partially to balance the pressure of the ground to reduce settlement. On account of the risks now known to result from working in compressed air (the 'bends', a temporary discomfort caused by nitrogen coming out of solution in the blood on emerging from a pressurised tunnel, may be controlled by observance of rules for decompression but the incidence of bone necrosis has been recognised as a longer-term and incapacitating disease) the use of pressures more than 100 kPa (1 bar) above atmospheric pressure is now avoided as far as possible. Up to 1 bar permanent effects are minimal, delays in 'locking-out' are avoided and, in consequence, where the tunnel is no more than, say, 20 m below the prevailing water-table, compressed air at such low pressure has on occasion been combined with pumping (Morgan *et al.* 1965).

Water inflows to a tunnel, possibly causing unstable ground, may be controlled by pumping, sometimes by vacuum-pumping at the face, accompanied by recharge where unacceptable ground movements as a result of settlement would otherwise occur, or where other deleterious effects of drying-out, such as the effects on elm piles or archaeological remains, need to be prevented. While the pumping has usually been confined to the immediate vicinity of the tunnel, the particular geology may favour the use of more extensive pumping from a permeable bed of ground beneath troublesome water-bearing silts or fine sands. This expedient was, for example, used for the Clyde Tunnel (Morgan *et al.* 1965) and in a different form constituted Project Moses for the Storebaelt Tunnel (Biggart and Sternath 1996).

As hinted above, special expedients are not mutually exclusive. The design of an optimal system, using one or more expedients, has needed to take account of many aspects of planning, of knowledge of the ground and of the features of construction (Kell 1963). The design of 'special expedients', where these are to be used, must form an integral part of the *design* process described in Chapter 2. Features affecting the present choice of excavation, lining and spoil disposal method are discussed in Chapter 6.

1.6 The developing problems of management

In the earliest days, tunnelling adopted traditional techniques, reliant upon the initiative of the miner to adapt to change. Each *qanat* for example was constructed by a small team, self-reliant and dependent on simple provisions by others of tools and materials. The Roman tunnels were usually built by slaves or by prisoners of war, requiring formal procedures for the organisation of vast bodies of men but providing little scope for initiative or innovation.

With the demands of an incipient industrial age, the practical techniques were developed in response to demand, with increasing understanding in a qualitative sense as to how to adapt methods of support, particularly of temporary support, to match the requirements of increasingly diverse varieties of the ground. Construction methods generally followed tradition. M.I. Brunel was the first to recognise a more complex relationship between design of the project and design of the means and the process of construction. Very conscientiously, but as a result subjecting himself to great responsibility, he took charge of the integrated process in preference to managing a system whose several components were the responsibility of different parties.

Then followed a period of increasing refinement in the planning of the tunnelling projects, avoiding difficult ground so far as technology allowed and designing the project to make best use of emerging technology and materials. The construction process remained fairly basic, apart from the introduction of tunnelling machines which were conceived to satisfy the requirements of the project designer as to dimensions, the tunnel constructor (usually the Contractor) as to performance. The Engineer, the project promoter's designer and contract administrator, was expected to be a competent designer who understood enough about the needs of the Employer and the problems of construction to be able to direct the project and to understand where the project design might be modified to assist in achieving objectives and in overcoming practical problems of construction. He was also expected to learn from experience as to how to achieve future economies. Many specialised techniques were developed in temporary works such as timbering and in overcoming specific problems but these solutions mostly arose from experience within the tunnel and hence communicated in practical terms between those concerned with the project, not requiring complexities in calculation or analysis.

As time passed, and we are now into the 1950s, there was an explosion in techniques and in the analytical ability to understand the criteria for success in the increasingly difficult types of ground being tunnelled. This, on the one hand, called for more specialised knowledge among those concerned and on the other for improved communication between those involved which was becoming, in consequence of growing complexity, increasingly difficult. The general increase in the amount of work was encouraging new entrants who lacked the continuity of experience. Tunnelling, largely for such reasons, began to be associated with uncontrollable costs and lengthening periods of construction. The profession of civil engineering should at this point have addressed this mounting problem. The problem was of a general nature but expressed most extremely in relation to tunnelling, on account of dependence on the ground and the increasingly specific requirements from management for success in the application of new methods and means of construction. The profession through this period stood clear of the problems of organisation and management, except in a remote, non-specific manner. The most debilitating feature was a fatal developing division between engineering and management and the increasing incoherence in the 'management' of the several elements of *design*. As a consequence, the initiative for action passed into the hands of other professions, the law and those concerned with accounting for the costs of projects. The results have been predictably unsuccessful with poor control of costs and dissatisfaction with the degree of achievement of the Owner's objectives. The time is now propitious for the civil engineering profession to recapture the initiative, impose order and achieve results for their Clients, taking encouragement from projects which have shown features of enlightenment in project procurement and direction described in following chapters. This is the general theme of this book. This Chapter is intended to set the introduction to the paradox that the tunnelling engineer has had the benefit of an increasing range of tools with which to address his tasks whereas the framework, in which such tools may be used to their best advantage, has generally deteriorated with time – with honorable exceptions. Engineering potential and management practice have been travelling in opposite directions with the consequence that new techniques may appear superficially and erroneously to have added to the problems instead of adding to the solutions. Subsequent chapters attempt to diagnose the issues more specifically. Solutions exist, with economies achievable by applying the best capabilities in all aspects of tunnelling and through an understanding of its successful integration, which lies at the heart of successful management.

Chapter 2

Design: the ubiquitous element

"When I use a word", *Humpty Dumpty said in rather a scornful tone,*
"it means just what I choose it to mean – neither more nor less".
Alice Through the Looking Glass, Lewis Carroll.

2.1 The nature of *design* and its application to tunnelling

Systems engineering, central to manufacturing industry and, in its widest sense, universal to all effective practical and political aspects of life, is a term little understood by civil engineers and hence is not commonly applied to tunnelling (Muir Wood 1996). It is important to understand the concept, however, in order to appreciate its potential even if we do not immediately start as tunnellers to use the expression 'systems engineering'. A more familiar term to the tunnel engineer is 'design' and if defined sufficiently broadly – as used in this book – '*design*' may be treated as synonymous to systems engineering.

Manufacturing industry differentiates between 'product design', i.e. the design of the finished article, and 'process design', the means for achieving the desired product. Commercial organisations with an instinct for survival have long realised the intimate relationship between the one and the other. As a very simple example, a chocolate cream depends on the use of an enzyme in its manufacture to liquefy the filling after application of the coating.

Design is the central element of the art of engineering and of architecture but members of the two professions tend to use the term in quite different senses. To many architects, and to many of those who designate themselves as 'designers', design may be related to questions of form or style of the artefact. Many engineers on the other hand use 'design' as a synonym for calculation and analysis of a finished product, related by civil engineers either to temporary or to permanent works. Fundamentally, however, in this book, *design* may be understood to describe the continuous thread, or at least what should be a continuous thread, of recorded processes of thought – with appropriate transferences to manual skills – in order that ideas may be successfully transformed into artefacts or projects, including essentially the processes of construction. At the heart of the *design* process is the concentration on the ultimate purpose of a project, to achieve the business objectives of the Owner, expressed in the broadest terms as including the interest of the public and of other stake-holders.

2.1.1 Characteristics of design

The characteristics of *design* will first be defined in general terms, then the process of *design* described, identifying particular features critical to successful *design* in tunnelling. It will subsequently be demonstrated that, once the centrality of *design* to successful tunnelling is accepted, funda-mental consequences follow affecting policy and practice.

The essential characteristics of the *design* process (Muir Wood and Duffy 1996) are that it is:

- *Creative*, demanding analysis as well as imagination, often applied in alternating sequences. Virtually all successful engineers and scientists operate in such a fashion; Poincaré describes the mathematician as working in this manner.
- *Holistic*, so that all relevant aspects of evolving solutions are taken into account. At the level of a total project, the aspects may be very broad, including viability, environmental and social consequences. As the project develops in greater detail, optimisation may well depend on revisiting some of the non-technical aspects of *design*.
- *Integrative*. Synthesis lies at the heart of the proficiency of the engi-neer. Synthesis implies the ability effectively to integrate contributions from those studying and developing different aspects of the same feature of the project.
- *Interactive*. Direct communication is necessary between members of a design team studying different aspects of *design*. A problem iden-tified by one participant may well be viewed as an innovative opportunity by another.
- *Iterative*, in recognition of the fact that, except in the simplest circum-stances, an optimal solution may evolve in interactive stages. This may entail trial-and-refinement or, occasionally, a totally fresh approach to a solution.
- *Cross-disciplinary*. It follows from the above that, for productive and progressive communication to occur, the principal participants in the *design* process need to combine a depth of knowledge in their own specialities with an awareness of, and respect for, the contributions to be made by other members of the team.

- *Systematic*, the work proceeding both in principle and in detail to a well-conceived and developing plan. The main skill in directing *design* is to combine the discipline of the plan of action with the flexibility essential to the *design* process. This procedure may be conceived as progressive exploration of only partially mapped territory as the mapping develops.
- *Ethically grounded*, recognising the interests of all 'stakeholders' and the need to respect social and environmental issues external and internal to the project, with clear lines of responsibility for taking such matters into account in reaching *design* conclusions.
- *Hierarchical*, with two-way flows between the levels of the hierarchy so that each contributes effectively to the evolution of the project. New issues identified at a higher level may be delegated for further study at a more appropriate level.

Design leading to successful projects is characterised by:

- the need for clearly stated objectives, including matters of resources, functional criteria and timing, agreed between the Client and the controller of the *design* process, referred to below as the 'conductor';
- performance criteria, modified and refined, in consultation, or in a tighter structure of management, with the Client as work proceeds, execution of the project and planning of its operation being seen as interactive aspects of the project (the term 'Client' being used here to represent the party commissioning the project, the 'Owner', or possibly his 'surrogate' representative, see Chapter 3);
- professionals engaged in a unified design team, such that discussions lead to improved understanding of the viewpoints and perceived objectives of other members, and hence towards most rapid convergence to the preferred scheme, through constructive dialogue;
- recognition that an appreciation of the potential risks is a necessary preliminary to the control of risk.

2.1.2 The parties to the design process

The Client owns the project and the ultimate criterion for success is that his objectives are achieved. The *design* process has therefore to be able to combine the consideration of operation and construction. By a sufficient understanding of each aspect, the objectives may become modified in a process of optimisation, as definition of the project develops. The implication is that there is a need for a clear statement of objectives, differentiating between those elements which are essential and those preferred but negotiable. Such a statement will provide a guide to the terms of reference for the contributions of each principal party to the achievement of the objectives.

The *design* process needs to be directed by a 'conductor', the term being deliberately chosen in preference to 'manager' since a manager has been too often cast in the role of undertaking a remote administrative function. The artificial division between engineering and management, fostered during recent years by mistaken notions of accountancy, has had a highly corrosive effect on the success of tunnel projects. Once the pervasiveness of *design* is understood, it will be realised that the function is highly pro-active, the 'conductor' engaged within the *design* process and effectively directing the activities of the protagonists. The *design* process, needing simultaneously to satisfy multiple criteria, will pass through stages of compromise, 'trading-off', towards the optimal solution. Usually the process will be one of trial-and-refinement, but sometimes it will be necessary or expedient to reshape the concept towards a radically different form, a likely condition for innovation. Thus, the *design* process entails the skilful aiming at moving targets, the rate of convergence upon its chosen option reflecting largely upon the cross-disciplinary skills of the 'conductor' and the 'players'.

Once the guiding principles for a specific item of the overall scheme have been adequately defined, the design of this item may proceed, with the 'boundary conditions' made known to all concerned, but periodically the several features of the scheme need to be reconsidered together to ensure that no new interactive problems have inadvertently been introduced.

Using the terminology of the system engineer, Figure 2.1 illustrates how *design* may be illustrated as a series of iterative loops, each loop entailing communication of ideas and knowledge between participants, possibly through the medium of computers.

Uncertainty is an issue in all engineering projects; tunnelling brings its own special brand of uncertainty, largely on account of the geological features, their impact on the method of construction and the impractibility of fully characterising an element of the ground even where, rarely, it is possible fully to explore it. Where information is contributed by any party to the *design* process, such information should be classified as to its reliability and precision. An important aspect of *design* will be the consideration of the means and justification of the cost of the reduction of uncertainty in a particular respect, where such reduction may bear upon project cost or risk.

2.1.3 Uncertainty and risk

Consideration of risk must be at the heart of the *design* process. Since *risk* and *hazard* are terms used loosely in different contexts with different meanings, each is defined, following BS 4778, in these terms:

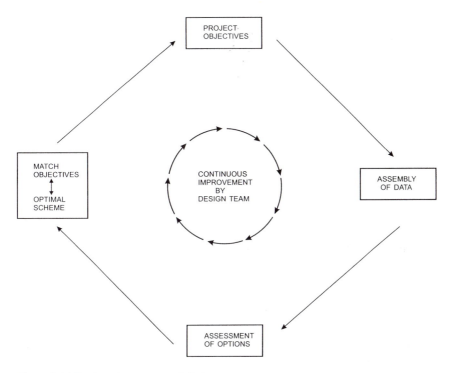

Figure 2.1 The iterative nature of design.

- *Hazard*: a situation that could occur during the lifetime of a product, system or plant that has the potential for human injury, damage to property, damage to the environment or economic loss. ('System' is here understood to include the construction phase of a project.)
- *Risk*: a combination of the probability, or frequency, of occurrence of a defined hazard and the magnitude of the consequences of the occurrence. (Since risk is related to a certain set of prior assumptions, it is logical that, while risk generally denotes a disagreeable outcome, it may exceptionally denote a favourable circumstance; this is a point made by the Institution of Civil Engineers and Faculty and Institute of Actuaries (1998)).

The driving force of good *design* in a tunnelling project is the intention to minimise undesirable risk (generally where the term 'risk' is used, this sense is implied) and to enhance the opportunities for the most appropriate, and often innovative, elements to the work. The features of *design* as defined in Section 2.1.1 are fully adapted to the investigation and control of risk.

The principles of risk analysis and risk control have been addressed by many authors (e.g. Royal Society 1992, Rimington 1993, Godfrey 1996). Health and Safety Commission (1991) provide a definition for the principle of ALARP (As Low As Reasonably Practical) for application to the establishment of reasonable limits to the reduction of risk.

Risk needs to be addressed on a project-wide basis. From the viewpoint of the Owner there may well be elements of the broader 'business case' for the project that need also to be included. Such considerations may take many forms, highly project specific, of which a few are set out below by way of example and reminder.

1. Optimisation of a single project may overlook features of standardisation which may bring benefits for several related projects, or which may better fit accepted administrative practices.
2. Training for operation, and encouragement of the optimal coordination between construction and operation, may require members of the Owner's team to be incorporated within the construction team during the later stages of construction.
3. The possibility of a changing future usage of the project may justify higher standards or more generous provisions by the project, e.g. a water supply tunnel which may in the future be operated under higher internal pressure.
4. Protection of a sensitive structure may justify stringent measures of design and control of an underground project, e.g. the Jubilee Line Extension at Westminster described in Section 5.3.
5. The terms of financing or of extracting revenue from the project may impose constraints on progress, with, for instance, an intermediate operating stage of partial completion.
6. It may be expedient to incorporate standards of health and safety higher than the prevailing norms, not only for reasons of principle but also to avoid retrospective variations as a result of new legislation during the period of construction.
7. Aspects of project risk, in cost and timing, may reflect on other features of the Owner's responsibilities.

The principal causes of risk in tunnelling may be expressed in these categories, summarised and developed from CIRIA Report 79 (CIRIA 1978):

1. The physical conditions of the work:
   • features of the ground and of ground-water interacting with the methods and means of construction;
   • presence of natural or man-made noxious or flammable gases or liquids;

- obstruction caused by pipes, services, wells, shafts, foundations or boreholes;
- unstable ground;
- seismic effects;
- presence of landfill.

2. Human failings:
   - changes in management structure or personnel;
   - failure to identify essential features of project management;
   - late access to site;
   - late issue of essential information;
   - examples of incompetence or inefficiency;
   - defective contract documents;
   - defects in design of the works, in the design of their construction or in design of the means of construction in relation to the possible range of conditions;
   - inadequate or inappropriate designations of responsibility;
   - defective workmanship.

3. Effects on other parties:
   - damage or injury caused by collapse or inundation of the works;
   - negligence in design or workmanship;
   - effects of ground movements on structures or services;
   - effects of changes to ground-water regime on structures or services;
   - consequential losses;
   - war or social unrest.

Many of these features are not specific to tunnelling but the particular characteristics of a tunnel – problems of access, degree of dependence on the ground, the linear process of construction as a bar to acceleration – enhance the consequences of such factors, individually or in combination.

The overall strategy concerning risk pervades all aspects of tunnel management but at the project design phase the elements of risk should begin to be defined and their containment related to aspects of project strategy driven by *design* considerations. The orderly control of risk may be described in these phases:

1. identification of potential risks;
2. investigation of nature and magnitude of potential risks;
3. development of measures to mitigate risk;
4. assessment of residual risk;
5. allocation of responsibility for accepted risk;
6. application of measures for risk control and remediation.

The essence of risk management is the integration across all aspects of the *design* process. Thus, for example, potential hazards revealed by site investigation in relation to a favoured method of construction may be eliminated in whole or in part by contributions from the planning or the design of the project.

The general rules for tunnelling set out first in CIRIA Report 79 (CIRIA 1979) – and subsequently in different form in many other documents as the benefits of risk analysis and risk sharing have become more widely appreciated – remain valid, namely that the first priority is to anticipate and forestall controllable causes of risk. This entails meticulous attention to procedural issues in the planning stage of a project in order to understand external constraints. Chapter 4 discusses the strategy for site investigation in order to identify potential hazards of the ground. Here it is appropriate once more to emphasise that so-called geological risks in tunnelling are represented by the product (Geological hazards) × (Susceptibility of scheme and means of tunnelling to such hazards).

The nature of tunnelling entails a degree of uncertainty. Risks may be foreseeable in nature but not in extent or precise occasion, and may not be practically capable of total elimination. Mitigation is best achieved by:

1. describing the nature of the hazards, their probable or possible incidence and extent in relation to different forms of tunnelling;
2. indicating the circumstances in which the hazards are likely to occur;
3. making provision for responsibility and for contingency planning should the risk occur;
4. devising and enacting inspection procedures, free from pressures for maximum progress or for cutting corners, to ensure that measures vital to control of risk are respected.

The second rule for risks, whether these concern costs or questions of health and safety, is that primary responsibility should be specifically allocated. The uncertainty that arises in the absence of such specific indication of responsibility has been the primary cause of failure to control much risk. Yet more costly and destructive of good management, is the practice, in place of risk identification, of the incidence of risk being simply 'passed down the line', usually to a contractor. Quality Assurance (QA) (see Section 7.3) has contributed to recent tunnel problems, where Permanent Work (the field of the Owner's QA) becomes divorced from Temporary Works in areas in which the combination of the two assure the safe construction of the project (Heathrow Express collapse [described in Chapter 9], Øresund Link, loss of tunnel element [see Section 8.2]).

Allocation of responsibility will take into account the ability physically to control the risk, where the immediate effects of the risk will be experienced and the ability to institute timely countermeasures. Beyond such general principles, the allocation of risk will be much influenced by the nature of the contractual relationships. Risk recognises no frontiers of professional discipline or demarcation of duties. A well-organised project will ensure that it is in the interest of all parties to identify and control risk, emphasising the particular features and the most expedient means of control. This is a practice to infuse the project, right from the outset, working from the general towards the particular as the features of the project assume definition and as some risks may disappear, others more precisely understood. The greatest benefits are achieved by modifications made before expenditure on the variation of any aspect of the project has already been incurred. A particularly expensive feature of the Channel Tunnel was that requirements of health and safety were evolving as the project was being designed and entailed changes after work of manufacture and construction was already in hand.

Statistics of past events may contribute towards estimating the risks to health and safety associated with particular forms of tunnelling but these should be treated with great reserve. Each project has unique features. The 'safety culture' of a specific project, as part of the general concern for good working conditions, will affect many of the factors which contribute to safe working practices. In particular, the extent to which there has been full exchange and publication of views between the Parties engaged on the project concerning the possibilities of particular risks will have a pronounced effect on their control.

Uncertainty about the ground is a hazard to be understood and, so far as possible, circumscribed. The geological model (Chapter 4) is based upon sampling of a very small fraction of the ground, on indirect interpretation of geophysical records and upon fitting the model into a wider geological context. Wherever an interpretation is made, alternative interpretations of the facts, possibly more or less favourable to different forms of tunnelling, should be mentioned. For a project to be constructed under competitive contract, and where there is uncertainty in one or more particular features of the ground which may have a large effect on the contractor's choice of method or of cost, the use of 'reference conditions' (CIRIA Report 79) indicate the assumptions to be made for the basis of the contract. The principle is stated in para. 4.3 of this Report:

'It is proposed that the Engineer, who will have had much greater opportunity to direct the site investigation carried out and weigh the results in relation to his design and the aims of the project, should define, within limits and where appropriate, the ground deemed to be foreseeable and so provide "reference conditions" for the ground. These would then be accepted, unless changed or modified by the tenderer, as the range of conditions which, in terms of Clause 12 of the ICE Form, would be used to judge if the physical conditions and artificial obstructions encountered. ... "could not reasonably have been foreseen by an experienced contractor". By doing so, the Engineer will have established a common basis upon which all parties can at the time of contract form an understanding of the physical conditions in which the work is to be performed.'

As set out above, the 'Reference Conditions' are related to the Institution of Civil Engineers (ICE) Conditions of Contract, up to the 6th Edition (ICE 1991) and, by analogy, to FIDIC Conditions. The principles, however, have more general application. The selected 'reference conditions' should have specific relevance to the particular project, defining features of vital importance to the methods of construction and their costs. As stated in CIRIA Report 79 para 4.6 'The ground encountered during construction should be monitored against the Reference Conditions.' This approach has two main objectives: (1) to achieve economy where there is a low probability of a major hazard (accepted by the Owner) which would, if charged to the responsibility of the contractor, affect the optimal method of working and in consequence the cost; (2) to provide a clear basis for measuring the extent of variation on account of the geology, effectively in so doing ensuring that the preparer of the contract documents has considered the nature of uncertainties with particular bearing on construction risks.

Too often, particularly where geological risk is imposed fully on the contractor, the Engineer claims, without justification – and in the absence of any, or any adequate, assessment of risk – after a particular geological problem has been encountered during construction, that he had foreseen precisely the difficulties that would arise to the contractor. The approach described above should assist in other constructive objectives of CIRIA Report 79, namely: (concerning the provision of site investigation information provided to the Contractor: para 3.4)

'The satisfactory solution is seen in providing the tenderers with the clearest possible description of the ground on which the tender is to be based and to expect in return a method statement indicating a full appreciation, based on careful examination and full understanding of this description.'

Uncertainty is pervasive. It may be desirable for a project with a simple purpose to define the requirements in terms of physical characteristics and performance. For a complex, innovative project, and for one with possibilities of interactions between different contractors, this is rarely achievable or even, if it were, the optimal solution. The wider consequences in contractual terms are discussed in Chapter 7.

One nearly ubiquitous hazard of urban projects concerns the presence of pre-existing service pipes and cables, either unrecorded by the relevant authority or misplaced sufficiently to cause embarrassment and delay in sinking access shafts. When in any doubt, it is wise to organise a physical or possibly a geophysical exploration to expose and then eliminate the problem. Boreholes are normally preceded by a trial pit for this reason. One such borehole in an open field outside the perimeter road of Heathrow Airport, in apparent open country and with an undisturbed agricultural air, omitted this precaution and struck oil – the insulation for a high tension power cable.

The nature of risk is such that care must always be taken in defining the assumptions on which estimates of uncertainty are based. For example, a sewerage tunnel project in England was known to face a risk of encountering bands of Devonian rocks of strength exceeding those penetrated by the site investigation. For the designed scheme of tunnelling by drill-and-blast, such a contingency would have only relatively minor effect on the overall cost of the project. An alternative scheme, offered by the Contractor and accepted for the project, entailed excavation by road-header, a scheme offering savings in cost of about 10% for the predicted geology but susceptible to high increase in cost should the bands of strong rock be encountered – as they were on account of misinterpretation of the geological structure. A differential risk analysis would have revealed this risk before the cost was incurred. A decision would then have been made either for supplementary site investigation to establish whether the tunnel would be liable to traverse the bands of strong rock in which case the excavation by road-header would be reassessed, or to reject the excavation by road-header at the outset. In the event, high wear and low productivity caused the road-header to be removed and excavation completed by drill-and-blast at an overall claimed increase of cost of nearly 50%.

2.1.4 Qualifications for the design team

The leader of the *design* operation has already been referred to as the 'conductor'. Indeed, his functions are comparable to those of the conductor of an orchestra in that he knows how to read the full score (i.e. a broad grasp of the *design* process to be undertaken, how this may be fulfilled, the pacing of the process). He also knows what contribution to expect from each instrument (i.e. from each key *player* in the *design* process). The 'players', for their part, understand their instruments and how these blend with the remainder of the orchestra. The overall 'intellectual profile' may be illustrated in the form of a 'wind-rose' (Figure 2.2) for a particular project with the several levels of capability of each participant defined as:

- *Expertise*: wide understanding in the particular area of contribution, in relation to the needs of the project.
- *Competence*: general knowledge in the areas of contribution of others adequate to allow engagement in constructive dialogue.
- *Awareness*: understanding adequate to respect the contributions of others in such areas.

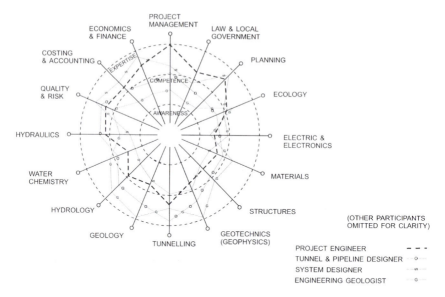

Figure 2.2 The of mix of skills needed for a water supply project.

While the *design* process needs a 'conductor' throughout, the occupant of this post may well change, particularly for a major project many years in its evolution, in consequence of the changing characteristics of the different phases of development of the project. Any such change must however respect the need for continuity in conceptual thinking. This is particularly necessary for tunnelling projects, since features of the process design, i.e. the design of the operations of construction, need to be fully appreciated from the outset in all aspects of planning and investigation.

2.2 Steps in the *design* process

Chapter 7 describes the inception of the underground project, at which time, the *design* process, in the absence of adequate understanding, is most likely to be inhibited. Where a tunnelling project has set off down a road of fragmentation of responsibilities and control by an administrator without understanding of the subtleties of *design*, radical measures are necessary to restore the conditions for an optimised project. Figure 2.3 gives an indication of some of the principal lines of interaction between the several activities. An overall programme needs to be developed, indicating specific occasions for policy decisions in the course of development, for which sound advice will be needed from those supervising the several activities. For example, a Parliamentary Bill or authorisation by a Planning Authority may impose constraints upon the features of the project, and

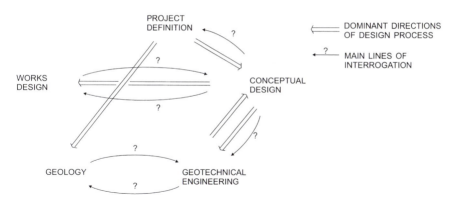

Figure 2.3 Principal interactions between participants.

may itself be a necessary preliminary to the generation of financial support for the project. For this purpose it will be necessary to describe the approximate geometry of the project which must assume a basic knowledge of the geology sufficient to provide guide-lines for a preferred route and limits of deviation. There may well be other constraints to be considered at this stage, including land ownership, protection of underground aquifers for water supply, environmental features, urban planning, possible associated development, which will affect such a decision.

For the Øresund link between Copenhagen and Malmo, an Illustrative Design of the project was prepared by the Project direction team of the Owner and Consultants, for negotiations with authorities, to establish estimates of cost and a feasible programme and to provide guidance – but not to usurp responsibility – to the design-and-build Contractors (Reed 1999). The discipline of so doing undoubtedly helped to prepare the Project team to ensure that a selected Tender addressed adequately the practical issues of the Project. Such procedures are discussed further in Chapter 7.

At the initial stage many features can only be expressed qualitatively while, apart from the exceptional circumstance of total confidence in knowledge of the geology, construction costs for different options can only be expressed within margins of uncertainty whose width is highly dependent on the particular project. At this point, and throughout the early stages of developing the concept for a tunnel project, the 'conductor' should indicate the degree of uncertainty in cost (and other attributes) with indication of the points in the evolution of the project at which phased improvements in precision may be expected as the result of increasing refinement in the project definition to be expected from contributions from the *design* process. This enables a strategy to be developed concerning the taking of fundamental decisions and on the timing of financial input.

Where there are options for different forms of construction, policy dictates the extent to which these should remain open as the project proceeds. Figure 2.4 indicates possible effects on project optimisation at stages in definition of the project dependent on the decision to retain two alternatives A and B. Scheme B is indicated as less tolerant to uncertainties in conditions of the ground than Scheme A, but offering potential savings in cost (and possible benefits in operation including maintenance). At each stage in defining the project, the question needs to be asked as to whether further studies (which may also help to reduce uncertainties concerning Scheme A) are justified in order to establish whether the potential benefits of Scheme B may be realisable. There comes a point at which either Scheme B is favoured as the major uncertainties have been resolved or, where the doubts about Scheme B would entail further studies unjustifiable in time and cost, and Scheme A will be favoured. After the point of decision, there may well be justification for further studies to help to refine the selected option. If the timing for decision is related to the receipt of alternative tenders for construction, the differential value of operational benefits must have been previously assessed (and preferably provided to Tenderers) in order to assist in making the choice.

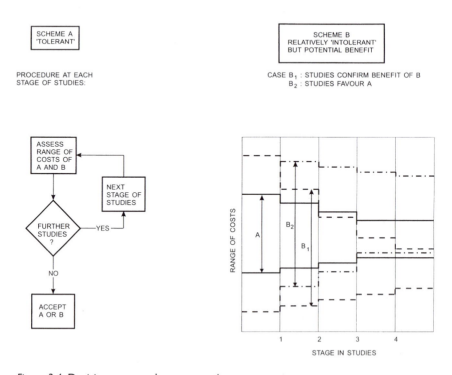

Figure 2.4 Decision strategy between options.

For example in the London clay Scheme A might represent a shield-driven tunnel with precast segmental lining, Scheme B could represent a tunnel excavated by road-header with primary support by sprayed concrete. Scheme A would tolerate a tighter layout than Scheme B, but a higher radius for curves. Scheme B might require additional site investigation to explore in greater detail the consequential effects of ground movement and particular features affecting face stability. Estimates of cost associated with specific risk factors at each decision point of project definition will indicate whether potential savings, through the flexibility in geometry offered by Scheme B, would justify additional costs of taking forward to the next stage both Options A and B. An important question affecting such a decision will include the expected degree of certainty in relation to costs, as a result of incomplete information about the behaviour of the ground.

As the *design* process proceeds, care should be taken to prepare and conserve records of all activities, especially those entailing investigations external to the project and dealings with other parties. The basis of an elementary catechism as an *aide-memoire* is illustrated by Table 2.1 (based on Anon. 1977), which may be developed in greater detail appropriate to the specific project. Several benefits follow from such a methodical practice, beyond the compliance with quality assessment and associated QA procedures (which are too often over-simple in concept, unreliable in control of risk and over-complex in administration):

1. The state of knowledge and expectation at a particular time is clear as the basis of reaching a particular contingent decision. The desirability of reviewing the decision may then be decided in the light of specific revisions of the basis on which it was made.
2. A hiatus in the definition process of a project or the sudden transfer of responsibility introduces, in the absence of methodical recording of progress, a risk of subsequently duplicating work when the project resumes or, far more serious, the omission of a particular function in checking an essential external feature in the mistaken supposition that it has already been undertaken and has revealed the absence of any problem.
3. The catechism serves as a reminder that absence of information does not permit the elimination of a particular risk without good cause.
4. In the event of subsequent criticism about a decision taken earlier in a project (this particularly by a lawyer in the event of litigation to whom, once the out-turn is known, correct prediction at an earlier time is declared to be obvious, regardless of the limited data available at the time) the rationale behind such a decision may be judged in the light of information available at the time it was made.

Table 2.1 Excerpt from Professional Performance Audit prepared within Halcrow in 1977 (Anon. 1977)

3.4 During the course of design, consideration should be given by the Project Engineer to the following specific points:

  (a) Has there been any difficulty in interpretation or departure from recommendation in Codes and Standards?
  (b) Have any new criteria, evolved for particular design problems, been scrutinised independently by an engineer experienced in the subject?
  (c) Has adequate provision been made in the design and detailing for accommodation of differential movements caused by ground settlement and temperature changes?
  (d) Have special problems of climate, pollution, aggressive soils and ground-water been adequately considered?
  (e) During construction is there any risk of imposing excessive lateral ground loads on piled foundations or retaining walls by adjacent earthworks?
  (f) Where computer aided design methods have been adopted, have the programs been adequately justified?
  (g) Are the design, construction methods and materials best suited to the country in which they are to be used? Are the proposed standards of finish practical and technically and aesthetically appropriate?
  (h) Is the general appearance suited to the environment?
  (i) Have suppliers' claims about the properties and durability of their products been adequately scrutinised, particularly for a new material where access for replacement is difficult or impossible? (Such a defect could result in an inbuilt failure mechanism.)
  (j) Does any aspect of the design present difficult maintenance problems?
  (k) Has the design been correctly translated on to the drawings and have they been checked? Are the textual contract documents compatible with the drawings? Do the documents draw attention to all special requirements relating to controlling and accepting design criteria and monitoring performance?
  (l) Does the design comply with all statutory requirements?
  (m) Have all of the interested authorities been consulted and any necessary formal consent obtained?
  (n) Are all of the wayleaves as may be necessary to give access on to the site satisfactorily in order?

---

5. Such documentation, which should be concerned more with the methodology than the precise quantitative nature of the conclusions, much assists the appraisal of the direction of a project.

In relation to 3. above, a form of words often found in contractual documents reads to the effect that 'If there is reason to suspect risk of. . . .' without guidance as to how such a risk may be safely dismissed. The water transfer scheme by tunnel between the Rivers Lune and Wyre suffered a serious methane explosion at the Abbeystead valve house in 1984 causing loss of life following completion of the scheme in 1979 (Health and Safety Executive 1985). It was claimed that there was no reason to suspect risk from methane concentration. Within about 7 km of the line of the tunnel, a borehole had been sunk in 1966, the records of which were lodged with the British Geological Survey as 'commercial in confidence'. A 'commercially confidential' borehole in Britain is likely to be prospecting for hydrocarbons. These circumstances are discussed in greater detail in Section 8.3.

At the early phase of project planning, a 'sieving' process needs to be established in order to establish whether *prima facie* circumstances exist for particular forms of geological hazard. Many may be dismissed as a result of the first 'coarse screening'; others may require a finer mesh, i.e. a more detailed examination, to establish whether the study phase should attempt to identify and quantify such residual potential hazards.

As discussed in Chapter 3, in earlier days a tunnel might be developed in a 'linear' manner, with the determination of the scheme of construction allowed to follow the previous end-on functions of planning $\Rightarrow$ investigation $\Rightarrow$ project design. The increasing diversity of construction methods, their greater dependence on particular characteristics of the ground, the availability of more, and more specific, special expedients, the ability to obtain, and to apply in a quantified manner, more detailed information about the ground, the consequent ability to tunnel through more difficult ground, the greater differential in cost between the favourable and unfavourable circumstances, all tend to set the design of construction in a more prominent – and interactive – position in the *design* process.

As project definition proceeds, so will the importance of the design of the construction process become increasingly evident. Different tentative conclusions may be reached for each stage of the *design* with regard to the significance of construction design:

1. that the scheme entails application of familiar technology in conditions that may with confidence be adequately defined to permit this part of the work to be clearly specified and undertaken by competitive tender limited to selected contractors with appropriate capability;
2. that the scheme entails innovation of a nature which requires particular features of the construction design to be designated by the engineer, knowledgeable in the appropriate aspects of tunnelling, responsible for the project design;
3. that the scheme depends on development of the construction scheme in such detail that a selected tunnelling contractor should be incorporated within the *design* team. The circumstances might include one or more of these possibilities: extending the range of previous experience; the use of special expedients coordinated with other processes of construction; the development of special plant; experimental work to establish feasibility; the reliance on observational techniques to determine the scheme of construction.

2.3 Examples of application of the principles

While there are many examples of successful adoption of procedure (1) of Section 2.2, using familiar methods in what are expected to be familiar circumstances, lack of success has usually followed from failure to identify particular hidden departures from expectation.

Heathrow Cargo Tunnel (Muir Wood and Gibb 1971) is an example of procedure (2) where a new form of expanded lining was to be used in circumstances which required a short shield with massive provision for face support and a speedy reliable facility for lining erection and stressing, described in greater detail in Section 5.2.3.

The several forms of construction by design-and-build such as BOOT (Build, Own, Operate, Transfer) offer opportunities for full integration of project and construction design, unconstrained by traditional practices in the Public Works sphere, but it is remarkable that too often the two functions are undertaken at arms length and the opportunity is lost. Many other forms of 'partnering' offer the chance to set construction design fully within the *design* process.

Enlightenment at the top of organisations embarking upon working in a cooperative manner with all the benefits that this should confer need to take action to ensure that the intention pervades deeply into the operational structure. Engineers are not by nature belligerent, but years of working in defensive postures may have developed problems to working relationships that need positively to be eradicated. Training needs to emphasise those features essential to good practice and those which may well be relaxed or modified for the overall benefit of the project.

Each tunnel has a purpose beyond the formation of a hole, or series of holes, in the ground. There is, as it were, a light of reason at the end of the tunnel! There is a special skill in knitting together the desired operational features and the means for their fulfilment. For a transport tunnel built for an existing authority with the experience of operation traffic through tunnels, considerable knowledge and experience will be available concerning the features needed to satisfy the objectives, but not necessarily in the same manner as pre-existing tunnels. From the viewpoint of construction, observation of the route for *design* described in Section 2.2 should ensure that the considerations for operation will be adequately represented within the design team, conducted by an engineer capable of orchestrating the contributions (Figure 6.1). It is necessary to establish adequate lines of communication, recognising that aspects of operation run right across all aspects of project design. It is then that the capabilities of Figure 2.3 are tested to establish that there is adequately shared appreciation of the possible variants in the means for arriving at the optimal scheme. Furthermore, such a degree of understanding should help to ensure that optimisation is achieved in terms of life-cost (i.e. in maximising revenue [if any] less the sum of the costs of capital, operation and maintenance [all expressed in comparable terms of Present Value]) and social benefit, and in preparation for subsequent phases of development.

2.4 Construction (Design and Management) Regulations

Construction (Design and Management) Regulations (CDM) are set out for the United Kingdom in a document, *Designing for Health and Safety*, from the Health and Safety Commission Construction Industry Advisory Committee (CONIAC 1994) of 31 May 1994 in response to EC Temporary and Mobile Construction Sites Directive (92/57/EEC). Apart from reminding the several parties to a construction contract of their duties in relation to safety, the CDM regulations impose upon the Client the duty of appointing a 'Planning Supervisor' (PS) and in designating the 'principal contractor' who has duties concerning the coordination of management of issues concerning health and safety which may affect other organisations working at the site.

The Planning Supervisor is charged with ensuring that principles are applied to avoid foreseeable risks, to combat risks at source and to give priority to general safety measures. The project designer is seen to have the greatest input to safety issues during preparation of the feasibility studies and the concept/outline design. The designer has a responsibility at the procurement stage in establishing the competence of the contractor but during the construction phase the contractor is entirely responsible for health and safety apart from design variations which need to be transmitted via the PS. The Health and Safety Plan (P) is prepared in two stages, before and after the appointment of the contractor. A Health and Safety File (F) is also prepared for the benefit of those concerned with the operation and maintenance of the project after completion.

The PS is responsible for preparing the Health and Safety Plan which, on appointment, is passed to the principal contractor for development. Warnings and emergency procedures for identified hazards need to be established and the PS is to be informed of unforeseen eventualities.

The effectiveness of such a scheme depends greatly upon the competence of the PS, including the practical understanding of tunnelling. Otherwise, the prospect exists of the elimination of promising innovations by the PS on account of unfamiliarity, lack of imagination or lack of willingness to accept responsibility. The procedure may positively, with advantage and familiarisation of all concerned, lead to a greater extent of demonstration trials where a particular promising innovation undergoes a full-scale trial prior to its adoption. The Victoria Line in London (Dunton *et al.* 1965) provides one of many examples (Lane 1975) of the benefit of such a practice. The penetrating observation of a previous

Director General of the Health and Safety Executive (Rimington 1993) needs to be borne constantly in mind: 'to pay too much attention to avert any harm is likely to increase harm in the long run.' A further weakness of the functions of the PS, which derives from the practices on which the system had to be imposed, is the discontinuity which occurs at the time of appointment of the Contractor. It is necessary to ensure that CDM does not inhibit the development of linkages between the several aspects of *design* discussed above, essential to its effective operation, particularly those aspects of tunnelling which desirably entail an element of observational design (Section 2.7).

2.5 Pitfalls in the *design* process

As already hinted earlier (Section 2.1), the greatest source of problems derives from the failure to integrate management with engineering, whereby the *design* process is stunted and the project potential unfulfilled. There are many examples of the 'management' of major projects being committed to management consultants who 'manage' by ensuring compliance to a detailed instruction book, without regard to the features of interaction essential to the well-being of the project. When things go wrong, the system will ensure that the blame falls on another party, but this has nothing to do with the optimal – or even the acceptable – standards for a project. The management consultant keeps his 'Teflon' image with the financial world.

Chapter 7 describes the essential elements of management in relation to *design*, avoiding these twin, and often coexistent, defects of fragmentation between the several elements of *design*, and of an artificial division between engineering and management. When it is not integrated into the engineering, project management equates to administration, a negative, non-productive and non-creative function.

In the initial stages of a project, while there may be recognition of the need for information on a variety of aspects, broadly covered under categories of 'planning', 'design' and 'studies', the common defect is failure to inter-relate these functions. As indicated by Figure 2.5, there are many different aspects of each category, with some of the interactions illustrated by Figure 2.3. In the absence of effective management of the interactions, which may desirably be of an iterative nature, not only can optimisation not occur but advice on certain aspects of the project may be based on totally false premises. This feature may be illustrated symbolically for a single step as below.

Effectively, a, b, c, ... represent the known state of each aspect at the start of the stage of the project, with the view of each being developed to an improved state, A, B, C, ... respectively during the current stage. If we use the code

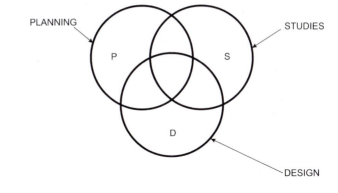

PLANNING

• FINANCIAL
• PROJECT
• LOGISTICS
• LAYOUT
• OPERATION

STUDIES

• GROUND
• DEMAND
• ACCESS
• AD HOC

DESIGN

• PERMANENT WORKS (PW)⎫ (PRODUCT)
• TEMPORARY WORKS (TW)⎭

• METHODS OF CONSTRUCTION⎫ (PROCESS)
• MEANS OF CONSTRUCTION ⎭

Figure 2.5 Elements of planning, design and studies entailing interaction.

$$x \Rightarrow X\{Y, z \ldots\}$$

as indicating that x advances to X based on a developed understanding of state Y and an undeveloped understanding of state z, our aim will be to reach

$$a \Rightarrow A\{B,C,..\}, \quad b \Rightarrow B\{A,C,..\}, \quad c \Rightarrow C\{A,B,..\}$$

indicating a full exchange of understanding. If however the aspects are developed end-on in the order a, b, c, ...,

$$a \Rightarrow A\{b,c,..\}, \quad b \Rightarrow B\{A[b,c,..],c\}, \quad c \Rightarrow C\{A[b,c,..],B[A(b,c,..),c]..\}$$

which is evidently far from a satisfactory state. If the features are developed in parallel but separately, then:

$$a \Rightarrow A\{b,c,..\}, \quad b \Rightarrow B\{a,c,..\}, \quad c \Rightarrow C\{a,b,..\}$$

an even less satisfactory conclusion.

In practical applications, this defect in lack of interaction can have serious consequences. As a simple example, the layout of an underground complex is planned on certain assumptions of the geological structure.

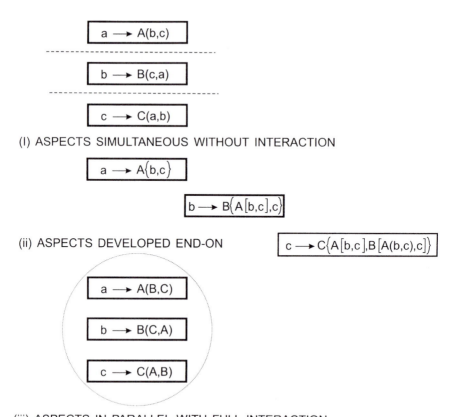

(I) ASPECTS SIMULTANEOUS WITHOUT INTERACTION

(ii) ASPECTS DEVELOPED END-ON

(iii) ASPECTS IN PARALLEL WITH FULL INTERACTION

Figure 2.6 Developments of aspects of a project.

A site investigation, commissioned on a predetermined basis, discloses unexpected features. The layout has assumed a certain method of construction, no longer viable against the new understanding of the geological structure. Each of these studies, undertaken independently, has in consequence arrived at irrelevant conclusions. In a comparable but less elementary example of such an incoherent approach, the project development proceeds in the absence of awareness of the weakness of its geological foundations which, when known, prompts a change of route. It is then too late to modify the investigation adequately to explore the new route. Geological sections need to be based on conjectural extrapolation. Each study should set out to achieve stated objectives on the basis of stated assumptions. Interactions will be aided by the understanding that any discovery of departure from these assumptions will be immediately reported in order to allow assessment of the significance of this change on the overall strategy and on the direction of parallel studies. Part of the skill in directing such studies is to perceive how best to advance one such aspect in relation to the progress of others in order to minimise repetition. Surprises are to be expected in the initial stages of the definition of a project and studies need in consequence to be commissioned in a sufficiently flexible manner to provide for modification in response to events. Irrelevance or inadequacy in any respect will otherwise cause problems and raise costs throughout the project.

Where the project develops through well managed and inter-related contributions, risk analysis, as discussed in Section 2.1.3, may readily be woven into the process. Risk identification has to be undertaken holistically since, as needs constantly to be stressed, risk is a consequence of association between contributory factors. Where aspects of a project are fragmented by separate uncoordinated commissioning of participants (often in competition, providing no opportunity for injection of proposals for improved coordination), it becomes unclear as to who, if anybody, is responsible for risk assessment. A massive site investigation was undertaken for a deep sewer project. For drill-and-blast tunnels with a precast segmental lining, the principal risk might have been that of supporting the tunnel through unstable ground associated with faults. In fact, the tunnels were specified to be constructed by TBM with *in situ* lining to follow subsequent to completion of excavation. The principal risk then became that of limiting water inflow, affecting not only the construction process but also threatening excessive settlement of buildings overhead. The site investigation had been inadequately directed towards this vital feature with the result of a thwarted contract. There was no evidence of systematic risk assessment having been conducted as to the adequacy of interpretation of the site investigation in relation to the options for construction.

Elementary deficiencies in design are introduced where the 'designer' of the finished tunnel fails to understand the problems and limitations of the means of construction. One such example is that of tunnels spaced too closely together for one or more tunnelling option. Another example arises from the inadequate provision of space for tunnel intersections discussed in Section 5.2.2.

Problems may arise from requirements for instrumentation within a tunnel which pay no regard to the practicalities of construction. One simple example arises from the wish to measure the tunnel diameter at a position where the operation will be obstructed by an impenetrable item of plant. Another example sets unguarded instruments exposed by projecting within the tunnel intrados, with virtually certain risk of damage. In the first instance, a solution will be found by precise surveying along the tunnel, by the measurement of tunnel chords or by the use of multiple head extension rods (MHXE) set in the ground, possibly ahead of the tunnel. In the second instance, the solution may be as simple as avoiding a particularly vulnerable level along the tunnel.

Each such simple, but actual, example reveals at least two contributory causes:

- lack of understanding and communication between participants;
- failure of management to provide the conditions to ensure such communication.

A general failing arises from misconceptions as to the nature of design for a design-and-build project. Thus project management calls for design proposals and receives a description of the finished features of the tunnel complex. Usually, the most important aspect of design will concern the intermediate stages towards the achievement of the finished tunnel and the means, sequence and timing whereby these are to be undertaken. In the absence of responsibility which crosses the boundary between the several aspects of *design*, failure is to be expected. In relation to the proposed use of NATM-type tunnelling in London clay, the following observation was made prior to the collapse of the Heathrow Express Rail link tunnels on 21 October 1994 (see also Chapter 9):

'Conceptual design and construction [of NATM] are particularly interdependent since the former may depend upon quite specific features of the latter for success, with the need to ensure that these are rigorously implemented. Present trends in commissioning tunnelling tend to ignore a condition for good tunnelling: the overall management of the design process. The many engineering activities of a project are subdivided and performed sequentially or separately, with only limited coordination. This ensures that interactions cannot occur and that the specific needs cannot be addressed in the early phases'

(Muir Wood 1994b).

Following the collapse, the management structure was radically changed (see Chapter 9).

2.6 The observational approach

Observation is a key attribute of the effective geotechnical and tunnel engineer. Observation implies the ability to use the senses, mostly but not exclusively by sight, to identify features of significance, or potential significance, to engineering decisions. Many circumstances of the ground or of the construction process are too complicated to describe with exactitude. Observation of characteristics and comparison with similar examples elsewhere explicitly or by personal judgment, leads to an understanding of the potential risks and opportunities.

The engineer needs, so far as possible, to be able to measure what he observes. The geotechnical information helps to define what is important for the tunnelling processes. For weak ground, for example, much evidence will be derived from samples and test results, the latter *in situ* for preference on account of delicacy of soil structure and fabric. For small strains, soils may be stiffer than for large strains, a phenomenon unlikely to be detected other than by in-situ tests.

For strong rocks, the several Rock Mass Classification (RMC) systems, intended by their authors as bases for tunnel support, provide initial guidance, although different for each system (Muir Wood 1993), on what to observe and how to record the results. Principally RMC relate to defects – or potential defects – of the rock mass and not to the inherent properties of the rock material. The raw data and observations of variability, coupled with reflections on their significance and comparison against experience in similar suites of rock, will usually provide more valuable information than the same data entered into the algorithm of the particular RMC.

For weak rocks, the contribution from RMC is more limited since behaviour of the rock will depend as much, or more, on the rock material as upon the discontinuities. Attempts to base support needs for weak rocks on RMC figures have been notably unsuccessful. The possibility of an approach for weak rocks as simple – even if inadequate – as that for strong rocks is illusory since the RMC systems disregard the *gestalt* aspects of the properties of the ground, i.e. the characteristics dependent on the combinative nature of the several itemised features rather than an unvaried algebraic association of factors. A preferable approach will be to identify the dominant factors for a particular rock type, then to map these features in a multi-dimensional manner (Figure 2.7) based on information from projects in similar rock types, thus identifying the bounds for satisfactory support with acceptable safety margins. In this way, instead of an algebraic function of rock and tunnel properties being presented in 2-D, the N selected significant properties of a particular tunnel in a particular suite of rocks will be presented in N dimensions. In the simplified example illustrated, the degree of support is related to RQD and to N_s (stability ratio), with potential ground-water problems, expressed as $k \times H$, projected in the third dimension.

Suppose, for example, that a tunnel in stratified mud-rocks displays initial stability around much of the excavation except where thin stratification is combined with bands of contrasted stiffness and where the bedding intersects the tunnel within a certain range of angles of incidence, for which a certain degree of support is needed. We have here four or five separate variables whose relationships may be represented as points on an (imaginary) 4–5 dimensional plot. For any confined range of these characteristics (Figure 2.7) we can observe the 'safe' and 'unsafe' regions on a plot of any two selected characteristics. An approach of this nature would assist in the planning of excavation and support procedures, always respecting a degree of 'fuzziness' in the vicinity of the safe/unsafe boundary.

Engineers and geologists have traditionally, as part of their training, developed the power of observation. Currently, much time and effort tends to be wasted in assembling prescribed data, often painstakingly acquired at the tunnel face, to enable calculation of a RMC algorithm which is then filed in a geological log book but not applied to serve any further purpose. Photographic or video records for posterity, zoning of specific characteristics in relation to previously recognised patterns as having engineering significance, coupled with notes on trends with the advance of the tunnel, would be of greater value and would assist in encouraging greater alertness on the part of the observer who will more readily appreciate the value of the observations in relation to the practical criteria for stability. In particular, the observer will react to unexpected features. These thoughts are further developed in Chapter 4.

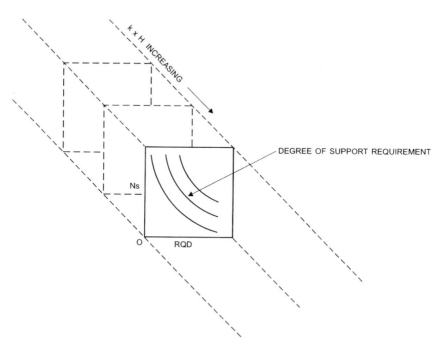

Figure 2.7 Mapping of properties of characteristics influencing support requirements in rock.

Observation is applied to decision-making in tunnelling in many different forms, which may be classified as follows.

1. *Informal*: action based on observation without intermediate assessment
   1.1 direct, e.g. incipient sign of face collapse.
   1.2 Indirect or inferential, e.g. identification of the trend of significant geological pattern at the advancing face associated with a potential type of problem.
2. *Formal*: observation associated with assessment
   2.1 inductive method, e.g. a reasoned formal basis for 1.2 above.
   2.2 design-based observation, the use of observation to improve the design base. For example, observation of the behaviour of tunnel support leading to improvement in design or construction techniques, e.g. achieving improved lateral stability of tunnel arches (see Section 5.2.1).
   2.3 Observational Method or Observational Design (as defined in Section 2.7).
3. *Syllogistic*:
   Application of model to syllogism: *a:b::c:d* where *a* represents analysis applied to a laboratory or other tunnel model, *b* represents observed behaviour of model, *c* represents analysis applied to prototype and *d* represents expected behaviour of prototype tunnel.
   [This is not generally considered as a specifically observational technique (any more than other forms of laboratory or *in situ* experiment) but where the model scale is large, or where a centrifuge model has been used to overcome dimensional factors or where the observations are based on test galleries or shafts, it probably should be.]

An example of 2.1 above might be where observation indicates cyclical variation of strata, such that the appearance of one geological member in the face will give warning of the presence of an associated weak or water-bearing member. Much interpretation of geophysical data is also conducted in this manner, with a known association of signal with feature at one point generalised towards inferring the same association elsewhere.

2.7 The Observational Method and Observational Design

Observational techniques for geotechnical engineering were recommended by Terzaghi and Peck (1967) under the designation of the 'Observational Procedure' and later formalised by Peck as the Observational Method (Peck 1969b). He recommended the following procedural steps:

1. Exploration sufficient to establish at least the general nature, pattern and properties of the deposits, but not necessarily in detail.
2. Assessment of the most probable conditions and the most unfavourable conceivable deviations from these conditions. In this assessment geology often plays a major role.
3. Establishment of the design based on a working hypothesis of behaviour anticipated under the most probable conditions.
4. Selection of quantities to be observed as construction proceeds and calculation of their anticipated values on the basis of the working hypothesis.
5. Calculation of values of the same quantities under the most unfavourable conditions compatible with the available data concerning the subsurface conditions.
6. Selection in advance of a course of action or modification of design for every foreseeable significant deviation of the observational findings from those predicted on the basis of the working hypothesis.
7. Measurement of quantities to be observed and evaluation of actual conditions.
8. Modification of design to suit actual conditions.

In principle, the adequacy of the first phase of a constructional procedure based on the Observational Method (OM) is checked by specified observations to establish the need, if any, for further work to achieve the desired objectives.

As set out above the procedures for the Observational Method are unnecessarily cumbersome, and often impossible to achieve in this form, for tunnelling. Moreover, statistical evidence of geotechnical variability for a tunnel could rarely be presented in a significantly reliable manner to permit the designation of 'most probable' (Step 3 above) condition. Tunnels may also lend themselves to 'zoning' whereby, for example, (Coats *et al.* 1982) expected qualities of the ground along the tunnel are characterised in relation to expected support requirements, with the intention of so designating each length as the tunnel advances. Where Zone 1, 2, 3 . . . are in descending order of rock quality and thus in ascending order of requirement of support, the expectation is that observation will require supplementary support to a relatively small proportion of lengths, originally classified as Zone 1, to the standard of Zone 2, with the comparable reclassification of some Zone 2 lengths to Zone 3. As a result of such considerations, Muir Wood (1987) recommended a simpler set of rules to apply to tunnelling, for a condition in which the need to modify the design might be expected to be exceptional:

1. Devise conceptual model.
2. Predict expected features for observation.

3. Observe and compare against 2.4. Are differences between 2 and 3 explained by values of parameters, inadequacy of 1 or inappropriateness of 1?
5. Devise revised conceptual model.
6. Repeat 2, 3, 4 and 5 as appropriate.

Such a procedure assumes that the design of the relevant feature, usually tunnel support, will have conformed to the conceptual model (1), with predesigned supplementary work undertaken where the differences between (2) and (3) so require.

In view of possible confusion, having regard to the wide acceptance of Peck's definition of the Observational Method (OM), simplified and varied approaches are collectively designated in this book as Observational Design (OD), emphasising that this is indeed a method of design. Eurocode 7 (EC7) (BSI 1995) includes the following remarks concerning an observational approach.

Because prediction of geotechnical behaviour is often difficult, it is sometimes appropriate to adopt the approach known as the 'observational method', in which the design is reviewed during construction. When this approach is used the following four requirements shall all be made before construction is started:

1. The limits of behaviour which are acceptable shall be established.
2. The range of behaviour shall be assessed and it shall be shown that there is an acceptable probability that the actual behaviour will be within the acceptable limits.
3. A plan of monitoring shall be devised which will reveal whether the actual behaviour lies within the acceptable limits. The monitoring shall make this clear at a sufficiently early stage; and with sufficiently short intervals to allow contingency actions to be undertaken successfully. The response time of the instruments and the procedures for analysing the results shall be sufficiently rapid in relation to the possible evolution of the system.
4. A plan of contingency actions shall be devised which may be adopted if the monitoring reveals behaviour outside acceptable limits.

During construction the monitoring shall be carried out as planned and additional or replacement monitoring shall be undertaken if this becomes necessary. The results of the monitoring shall be assessed at appropriate stages and the planned contingency actions shall be put in operation if this becomes necessary.

The organisational procedure for a project which entails an element of Observational Design is illustrated by Figure 2.8. There must always be emphasis on the time element to enable supplementary measures

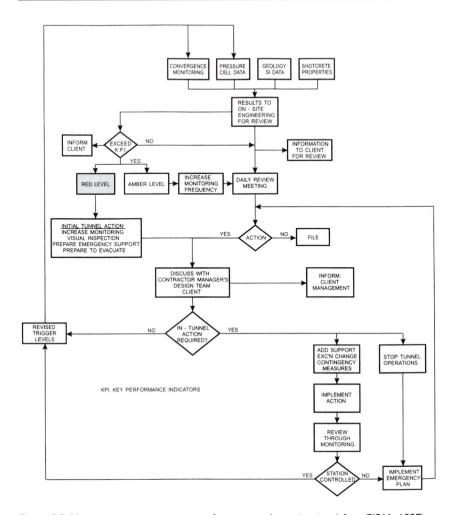

Figure 2.8 Management review process for in-tunnel monitoring (after CIRIA 1997).

(contingency actions) to be put in place, including all administrative actions, necessitating a rapid response system which may be a variation to normal management procedures of reporting and decision making.

The notion of the 'most probable conditions' of Peck, set out above, presents problems since it would imply that a high proportion of the work to which the OM is to be applied would require modified design and additional work in order to suit the *actual conditions*. Often, 'most probable' is interpreted as 'unlikely to be exceeded' so that modification to the original design becomes exceptional. In so doing, the engineer will be aware as to the nature of a failure mechanism which may allow a degree of dependence on mean conditions of the ground in the third dimension for a two-dimensional approach to design (see, for example, Appendix 5E). There is a rational basis for estimating the initial provision using OD, for greatest economy. This is illustrated by Figure 2.9 for an over-simplified example of the support requirement for a tunnel. Suppose that the requirement for support along a tunnel is established on the basis of a probability curve. Thus, the total area under the curve is unity, while the area under the curve between the origin and a vertical line *AB*, set at a distance *T* along the abscissa, is *p*, indicating that the provision of an amount of support designated as *T* has a probability of *p* of being adequate. Where this provision is shown by OD to be inadequate, the need for supplementary support thus carries a probability $(1 - p)$. If we can express the costs per unit length of tunnel of the initial support and of the supplementary support by the functions $f(p)$ and $g(p)$ respectively, the mean cost per unit length is given by:

$$C = f(p) + (1 - p)g(p) \qquad (2.1)$$

Thus,

$$C' = f'(p) + (1 - p)g'(p) - g(p) \qquad (2.2)$$

where the prime (') indicates d/dp.

For minimum cost, $C' = 0$ in eqn (2.2) and

$$g(p) - f'(p) - (1 - p)g'(p) = 0 \qquad (2.3)$$

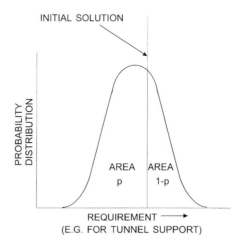

INITIAL SOLUTION UNIT COST: f(p)
SUPPLEMENTARY UNIT COST: g(p)
SO MEAN COST: f(p) + (1-p) g(p)

OPTIMAL VALUE OF p WHERE:
f'(p) + (1-p) g'(p) - g(p) = 0

INITIAL SOLUTION

PROBABILITY DISTRIBUTION

AREA p   AREA 1-p

REQUIREMENT ⟶
(E.G. FOR TUNNEL SUPPORT)

Figure 2.9 Procedure for applying Observational Design.

Hence, by knowing the variation in unit cost of initial and supplementary support in relation to the choice of T, the optimal value of p may be determined. By way of a very simple example, suppose $f(p) = Ap$ and $g(p) = kA(1 - p)$, A and k being constants, then $f'(p) = A$ and $g'(p) = -kA$. The value of p for minimum cost is then given by eqn (2.3) as:

$$kA(1 - p) - A + kA(1 - p) = 0, \text{ i.e. } p = (1 - 1/2k) \tag{2.4}$$

However approximate may be the basis for estimating the variability in the requirement, there is, by such an approach, a more rational and economic basis for establishing the initial requirement than that of determining the 'most probable' conditions of OM. The important feature is that of the ratio k between the unit costs of supplementary and initial requirement. In practice, of course, k and A will vary with p but it will be certain that $k > 1$. This simply reflects the fact that the ratio of unit costs between initial provision of support, as part of a routine, and of subsequent modification out of sequence will be less than unity:
    i.e.

$$g(p)/f(p) > 1 \tag{2.5}$$

In consequence, the value of p that should be selected for the initial stage of design will always be for:

$$p > 0.5 \tag{2.6}$$

and hence, except for a very skewed relationship of the probability of support needs, the initial provision should be greater than that corresponding to Peck's 'most probable' condition. In fact, as described above, the need for supplementary support may, on economic or safety grounds, be selected to be exceptional. The approach to design remains nevertheless that of OM or OD, regardless of whether or not the supplement is routine, occasional or quite exceptional.

    Some typical applications of OD to tunnelling are described in Chapter 5. Of course, OD has a potential breadth of application well beyond geotechnics, including instances in which the procedures are already being adopted without conscious awareness of the underlying principles. This occurs for example where a feature of a project is designed to be varied in relation to perceptions of changing demand. The system for application of OD is illustrated by Figure 2.10.

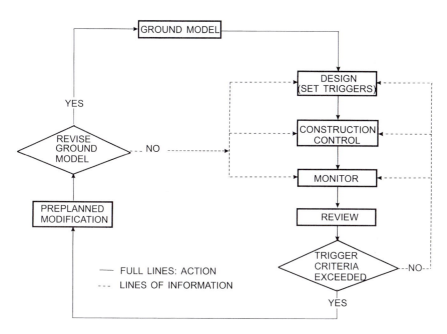

Figure 2.10 System for Observational Design.

Chapter 3

Planning

Expatiate free o'er all the scene of man;
A mighty maze! but not without a plan.
An Essay on Man: Alexander Pope.

3.1 Introduction to planning

Planning needs to be undertaken in a systematic manner, implying that it is a purpose-driven process, constantly concerned with priorities and constraints towards achieving the identified objectives. Throughout, there needs to be recognition of potential problems and their means of resolution, also of innovative features that may bring benefits to the project, as a result of identifying and resolving the problems. The project may depend on positive support from beyond the immediate project group, the Board for a private venture, a Committee or Local, National or International Legislature for a project with features affecting the public interest. The planning process needs therefore to take account of perceptions concerning the project and how these may be influenced by exposure to positive factual data. Misconceptions, once expressed, are often difficult to counter positively. Planning is undertaken in an iterative manner with periodical review of the objectives as their satisfaction becomes progressively clearer.

Economic choice may be made on the basis of sets of figures which should be fully comparable on the grounds of 'utility', i.e. suitability for specific purpose or purposes. Subjective issues will include engineering judgments as to practicability in prevailing degrees of uncertainty. Social and environmental features have a particular nature since each will entail a degree of irreversibility. A major project will affect demography, possibly in complicated fashions, for better or for worse. Direct environmental effects of a project – and of competitor projects which would be more damaging than the underground option – may be listed fairly objectively in the form of an Environmental Impact Analysis (EIA). The indirect environmental effects become more difficult to evaluate since the ecological interdependencies of species may be only partially understood, also the degree of success of adaptation to a changed set of circumstances. There is plenty of scope here for the 'pros' and the 'antis' to propose different scenarios and models for the interpretation of data. Development is generally hostile to natural life so that, as development in a region proceeds, so do the effects on the environment assume greater significance. Destructive effects are usually irreversible or only capable of being reversed by the injection of considerable investment.

Such thoughts lead to questions concerning the definition of sustainability (Muir Wood 1978). Sustainability is a global concept (Institution of Civil Engineers 1996) which needs to be considered as such in environmental terms in the way that resources are assessed for the 'use and convenience of man'.

Policy in relation to sustainability is not made for an individual project. Those concerned with planning individual projects need however to understand the features of the environment and of sustainability which may impact on the planning process. They also need to assess public attitudes and perceptions, which will affect the emphasis on different factors placed by those who may influence the success of the planning process. The planning process in Britain needs to be radically overhauled in order to be a help rather than hindrance to the selection of projects of development which contribute to the overall achievement of objectives in the public interest – but this is the subject for another book!

Planning has many connotations. At the most abstract, the evolution of types of solution to a stated problem or requirement is implied; at the most concrete, the establishment of a project layout in 3-D, with the organisation of resources for its achievement. Intermediately are to be found the justification for the choice, introducing considerations of technical, environmental, social, legal and often even political features (Figure 3.1). Each stage of planning has no precise definition but merges into preceding and succeeding stages. At each stage, success depends upon the ability to interrelate the several different considerations in order to enable smooth progress to the next stage. Planning is essentially therefore a multi-disciplinary, progressive process. It should not need emphasis that each aspect merits an appropriate and objective level of skill. Superficial assessment by those unfamiliar with the 'multi-dimensional' form of activity may well jeopardise a promising scheme, or occasion wasted effort in a wholly unviable project.

Tunnelling is not essentially different in its requirements for planning from other construction projects but underground planning does introduce a number of features specific or predominantly associated with working underground. These may be listed as:

- the liberty of choice introduced by flexibility in the third dimension;
- the extent to which cost depends on geology;

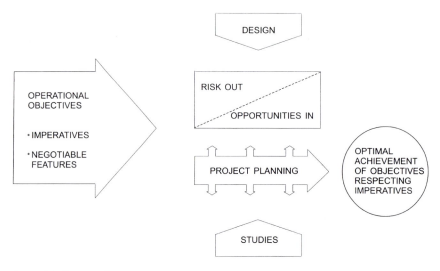

Figure 3.1 Project planning process.

- the many different benefits which may flow from freeing the surface;
- legal issues concerning the sub-surface;
- the (usually positive) environmental factors.

The technical input to planning is described in more detail in other chapters devoted to each specific aspect. Here, the emphasis is given to other 'non-technical' elements and to the overall process of planning. Those planning major projects, unfamiliar with work beneath the surface, tend in the early stages to view the tunnelling element as incorrigibly costly, but not to be subjected to informed scrutiny. A 'systems' approach to planning should indicate that, on the contrary, exploration of how to minimise the cost of tunnelling should assume a high priority in the overall planning process, since the high elements of cost are those most likely to produce significant reductions as a result of deepening studies of the options. Too often this logic is ignored until too late, by which time irreversible commitment to the overall 'shape' of the project prevents choice of a more satisfactory tunnelling element with consequent detriment to the project as a whole. Thereby the notion of unavoidably high costs of tunnelling is enhanced and perpetuated.

3.1.1 Assessment between options

The most obvious reason for a tunnel is to traverse a physical barrier such as a mountain range, strait, fjord or river. Rugged terrains, or areas subject to avalanche, heavy snow falls, landslides, floods or earthquakes may also favour a subterranean solution.

Again, tunnels may be preferred in open country for environmental reasons for the protection of areas of special cultural value or ecological interest. The environmental considerations in towns for traffic tunnels may include limitations of noise, of pollution, and of visual intrusion, the conservation of districts or even individual buildings of special merit, avoidance of severance of a community or to enhance surface land values.

In evaluating different responses to a specific need or requirement by surface, elevated or underground solution, the respective merits may be considered under these categories:

1. *Internal finance*: prime cost, financing costs, maintenance and operational costs, renewal costs, all set against revenue (if any);
2. *External costed benefits*: the value of the facility in terms of savings to direct and social costs external to the project;
3. *External uncostable benefits*: conservation, ecology, uncostable social benefits;
4. *Enabling aspects*: The project evaluated as a requisite facilitator of other desirable developments.

Among the important urban benefits for holistic planning (i.e. where adequate linkage exists between all those elements of infrastructure, including for example education and leisure facilities, in the planning of an urban region) but difficult to evaluate are environmental features which may influence the competitive place of the city in attracting industry, commerce, tourists and investment. These issues are only partially predictable, but represent marks of good design which stamp the identity of the city. Ray (1998) describes the part to be played by underground planning and construction in the UN initiative of the 'habitat' prospect. Of course, there are comparable considerations for tunnels in the country or to safeguard valuable archaeological sites.

In evaluating cost/benefit, one aspect that should be more frequently addressed concerns the separate identification of the groups who stand to gain or to lose from a particular project. For example, a road tunnel may provide general enhancement for the city, improvements in accessibility for those who live outside, all at the expense of a degree of congestion in the vicinity of the portals (Figure 3.2) and hence for such a project it is possible to identify the areas occupied by 'gainers' and 'losers' in relation to where they live and work. For many tunnel projects, the 'externalities', i.e. the benefits less the costs external to the project, will be found to be well in excess of the revenue.

The use of a simple 'yes' or 'no' referendum to canvass opinions on a specific project is always unsatisfactory (as it is in relation to any but the

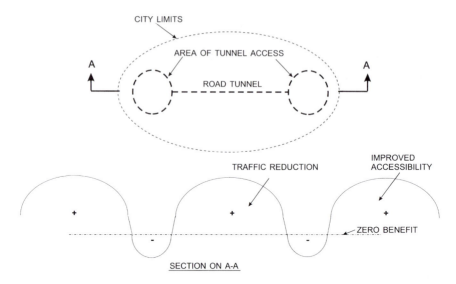

Figure 3.2 Social benefits: winners and losers.

simplest political decision). The strong opinions of those immediately affected or of those with coherent views on policy, into which a particular issue fits (or does not fit), are swamped by weak opinions, inspired by one-sided media or by irrelevant prejudice against a particular 'party' supporting or opposing the object of the referendum. Much success has been achieved by more transparent discussions on projects in their early stages of planning, in different forms of 'Public Participation', where the thinking behind a project is displayed for comment and discussion. Such exercises are not only of value in helping people to form informed views but are educative for the engineers and others concerned in effective two-way communication and in understanding issues which may not otherwise become apparent in the formal process of project planning.

3.1.2 In the beginning

At one extreme, an underground project may represent an extension of an operating underground system, e.g. for transport, water or communication. At the other, the project may arise from the public perception of a need to protect an existing surface facility or to provide an entirely new underground link. In the first instance, the planning process is a familiar continuation of what has gone before. In the second, a client needs to be identified and unfamiliar procedures set in hand. Where politics intervene, the situation may be likened to the Pirandello play *Six Characters in*

Search of an Author, a production played by many different casts to many different audiences, for example, for the Channel Tunnel over the years.

An essential element of planning for an unfamiliar project is to minimise financial exposure consonant with adequate thoroughness of each intermediate planning stage until the time of commitment to the project. At each stage, a tableau should have been set which allows the assumptions, upon which commitment is based, to be fulfilled. The planning of the decision-making framework is therefore an essential element in tunnelling, in view of the possible hidden traps. The planning process needs itself to be planned by coordination with the phasing of the supporting studies and preliminary design (Figure 3.3), which are mutually supportive.

For the *de novo* project, planning starts from a desk study which uses existing and readily accessible data on the nature of the demand, the possible types of solution and necessarily broad envelopes of the cost and time for achievement by each type of option. Unless the project is then seen to be non-viable on the most favourable assumptions, the scene is set for the next phase – and so for the subsequent evolution of the several options (Figure 3.1).

The initial planning phase is beset with problems, especially for the unwary. Options are wide open and specific experience untargeted in consequence, alongside a frequent lack of appreciation of the importance

PLANNING

- OPERATIONAL
- PROJECT
- FINANCIAL
- LOGISTICS
- PHYSICAL

STUDIES

- GROUND
- DEMAND
- MARKET
- ACCESS
- AD HOC

DESIGN

- PERMANENT WORKS ⎫ (PRODUCT)
- TEMPORARY WORKS ⎭

- MEANS OF CONSTRUCTION ⎫ (PROCESS)
- MANNER OF CONSTRUCTION ⎭

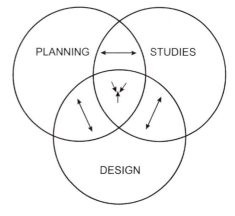

Figure 3.3 Integrated project planning.

Table 3.1 Incentives for tunnelling

1. Physical obstacles such as (a) mountains and (b) water barriers	(a) Road, rail, water transport (b) Road, rail, pedestrians, pipes, cables, conveyance of materials.
2. Urban planning and environmental protection	Sub-surface options for transport, services, communications through urban and sensitive areas.
3. Environmental control of temperature, humidity, noise, vibration	Wine, cheese, grain etc., shopping centres, laboratories, high precision industries, sheltered housing.
4. Economy and security	Hydropower, pipelines, sewerage and sewage treatment, cold stores, sports arenas, theatres etc., catacombs and burial chambers, military (offensive and defensive), reuse of disused mines, toxic and nuclear waste.
5. Extraction	Mining for hydrocarbons, metals, stone, salt, water supply.

of this initial phase. Experience in general construction management will probably be available with temptation to the innocent to derive cost estimates of a generalised nature – per unit area or volume, as for simple buildings, or by historic costs – to the underground options. This will not do; estimates of cost for each option need to be based on a rationale and applied in association with confidence limits, which must also have a sound basis for their expression, factors which must remain available to enable others to accept or qualify their reliability.

The diversity of purposes for underground construction prevents generalisation concerning the initial planning 'ordeal' which a particular option must survive. Table 3.1 sets out a list of purposes, sub-divided into categories. Even within a category, generalisation is limited, mainly in relation to the degree of constraint. For example, the continuation of a Metro or scheme of gravity drainage has limited options as to route and to level; caverns for oil storage will be located predominantly in relation to geology and access.

Another form of differentiation may be more productive, by classification through the form of client. The term 'client' is defined to mean the agency responsible for commissioning the planning of the project. There are three criteria for classification of clients, namely experience in the objective of the project and familiarity with tunnelling as an expedient to achieve this objective:

1. *The experienced client familiar with recent tunnelling.* Within this category one would expect to find the Urban Transport Authorities

who already operate an extending underground railway (referred to subsequently as a 'Metro'), those concerned with urban services of water, sewerage and possibly power and communication networks. Even where there has been recent experience in administration of a tunnelling contract, this does not however necessarily imply a familiarity with the essentials for good practice or the freedom to engage in appropriate methods of project procurement.

2. *The experienced client without familiarity with recent tunnelling.* This category includes established public and private providers of the service which normally adopts surface or above surface options or who have no recent experience of tunnelling. There is a gradation here which takes account of the degree of understanding of the opportunities and constraints associated with the subsurface option.

3. *The inexperienced client without familiarity with recent tunnelling.* Within this category will be found organisations with another principal function or those set up to undertake specific projects. While recruits will undoubtedly bring some experience and familiarity with tunnelling, in the absence of special foresight these attributes are unlikely to be tapped and channelled sufficiently early and appropriately in the planning stage of the project.

3.1.3 Planning unfolds

Each step forward in the planning process aims towards particular targets of establishing feasibility and in refining estimates by reducing the levels of uncertainty in relation to cost, acceptability and means of financing.

First, it is necessary to establish the factors considered *a priori* to influence acceptability of an option and its cost overall. As the planning and study process develop so may certain factors be eliminated, refined or combined by the acquisition and processing of data. There may well be problems in the association of factors, for example the state of the economy overall will not only affect revenue earning but also costs of construction. Effects of inflation need to be kept quite separate from monetary and fiscal effects of the state of the economy. The former is neutral, except in relation to the cost of borrowing money; the latter have far more complex consequences for demand and for cost of construction. Each planning scenario that is considered must eliminate incompatible features or the limits of the project's viability may be grossly over- or underestimated. This may be achieved by breaking down such factors, identifying the common element of each to be used in each algorithm for defining a scenario. (Figure 3.4) (A scenario represents one model among several for assessment of a project, based on a certain set of stated facts and assumptions).

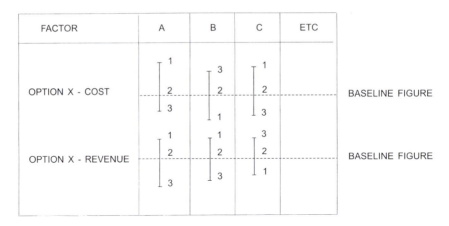

FACTOR	A	B	C	ETC	
OPTION X - COST	1 2 3	3 2 1	1 2 3		BASELINE FIGURE
OPTION X - REVENUE	1 2 3	1 2 3	3 2 1		BASELINE FIGURE

1, 2, 3....... SCENARIO NUMBERS
(SCENARIO 2 TAKEN AS BASELINE CASE)

Figure 3.4 Factorial contributions to uncertainty.

Many factors will be represented by mean values with a judgmental distribution which allows estimates of standard deviation. Tunnelling costs may well start from an assumption of uniform probability of cost between limits (Figure 3.5). As knowledge of the several factors advances, so may the limits of uncertainty be reduced with the uncertainty of cost at any time being represented as a curve of skewed shape as shown on Figure 3.6, whose contributory elements may be related to individual factors or to the combination of particular factors. One object of constructing curves of this nature, with their individual elements of uncertainty identified, is to provide guidance as to how best to apply resources during the planning process in order to reduce uncertainty in cost. Certain external causes of uncertainty may however increase with time. The compounded uncertainty curve of construction cost may pass through several phases of adjustment, affected principally by acquisition of geological information and the related assessment of schemes of construction, while that of demand will depend on the acquisition of data largely specific to the particular project. As this process proceeds so may the superiority of particular options become evident, comparison between options being aided by disregard of common elements of each. There may well be surprises when, for example, investigation reveals some previously unsuspected adverse feature of the ground.

Coordination of planning between projects for different purposes and serial planning in relation to succeeding projects may enter as factors for consideration (see Section 3.5). As emphasised throughout this book, each aspect of planning needs to be advanced in adequate association with

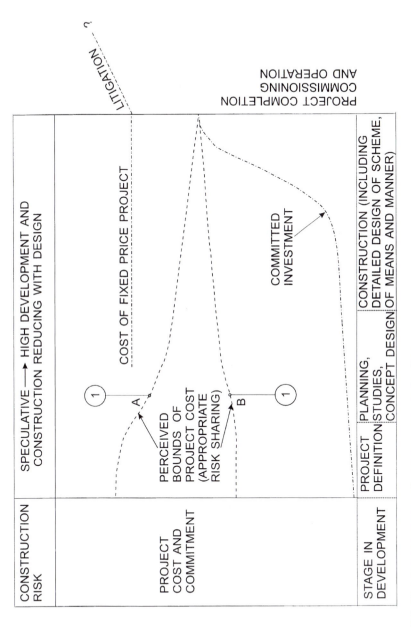

Figure 3.5 Changing perceptions of cost as a project develops.

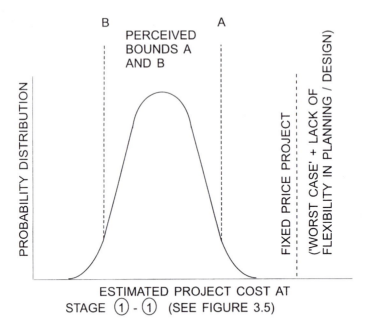

Figure 3.6 Project cost distribution.

other elements. Consideration of one aspect may introduce new opportunities or new problems to other aspects, with such linkages becoming more apparent with experience. The process proceeds in effect along the lines of American football with episodes of agreed teamwork interspersed with a huddle to agree the strategy for the next episode.

In terms of overall strategy, the objective must always be to tackle issues seen to have a dominant effect on viability. Interim planning reports will be required for a major project, grafted into the process so as not to interrupt the main elements of work. Certain aspects of the planning studies will be inherently costly and it must be a matter of judgment as to whether a particular aspect, e.g. of site investigation, should be taken rather beyond what is needed for a particular phase to avoid the additional, greater, costs of subsequent revisiting of this element. Assumptions made at the planning stage should be declared and recorded since, as explained in Section 3.4, if not so qualified the planning assumptions may be impaired. Table 2.1 indicates a phase in such a process.

Underground space has been used over the years for many forms of storage (and it is no coincidence that some of the finest wines and cheese come from regions of natural caves). More recently, the sub-surface has been used for offices, industry and sports facilities, also for other uses where benefit derives from an even temperature, particularly in regions of severe climate, from absence of surface induced noise, dust and vibration. Much study has been undertaken on the psychological effects of working underground and the features to reduce any element of claustrophobia. Some of the most innovative projects have combined different but complementary usages, e.g. Metro stations with underground car parks, tunnels for multiple services. The journal *Tunnelling and Underground Space Technology* addresses the many aspects of underground planning.

3.2 Financial planning

Tunnelling projects are constructed for long life, often assessed conservatively as 100–150 years. This feature will receive little credit from the rules of accountancy which discount future values (Institution of Civil Engineers 1969) by the formula:

$$£a_0 = £a_n/(1 + r)^n$$

where $£a_0$ represents present value, $£a_n$ value in n years time at a discount rate of r per cent.

There are essentially three periods relevant to financial planning (Figure 3.5):

1. preparation;
2. construction;
3. operation.

During period 1, costs will increase to a small percentage of the value of the project, while risk will reduce from its initial 'speculative' level. During period 2, the major expenditure occurs with outstanding construction risk gradually reduced towards zero on completion, or soon thereafter for the consequences of construction. During period 3, costs are recovered, in revenue or notionally. For BOT (Build, Operate and Transfer) projects, the period for deriving revenue from operation will terminate at the date of transfer (to state ownership). The sponsors of privately funded projects need therefore to recognise that equity investment raised for period 1, bearing high risk – and it is the perception of risk to the potential investor which matters – should be on a different basis from that for period 2. Similar curves may be constructed for revenue of commercial projects as the planning develops. A desirable feature in financing a tunnel project is to aim to replace the high risk, and hence high earning, capital of the early phases with share-holding of the completed project which reflects a more traditional commercial level of risk.

For infrastructure projects under PFI (Private Finance Initiative), there is a good case for period 1 for a 'basket' of projects to be borne by Government, with costs reimbursed during period 2 for those projects which proceed, with 'novation' of those at the centre of project planning to ensure continuity in thinking and planning. The reasoning behind this statement may be set out thus:

1. If the introduction of private finance is to be coupled with responsibility with the viability of the project, the timing for accepting this responsibility should not be too early or the element of risk to be borne (Figure 3.5) will make the price to the public purse excessive.
2. There must nevertheless be confidence that the studies and planning undertaken prior to introduction of private finance are adequate to reduce the elements of uncertainty in relation to the specific scheme of construction to be adopted.
3. Separate studies and planning by competitors at their cost in order to arrive at a condition suitable for engaging private finance would be burdensome and highly uneconomic.
4. Some potential projects may well be shown to be non-viable by the initial planning and studies.

During the course of negotiation for a specific project to be built under PFI, there may be circumstances (e.g. a project with few external benefits) in which the cost of planning and studies up to this time should be charged to the private participant, possibly even with an additional charge to represent a share of the comparable preliminary work for those non-viable projects which did not proceed.

The success of the investment plan must depend vitally on the quality of risk assessment and mitigation throughout the project, a feature emphasised throughout this book.

3.3 The law: facilitator or tripwire?

For any project, one of the first steps must be to establish the legal environment in relation to the project, since the law may, and it usually does, come to occupy the critical path and is remarkably resistant to stimuli for acceleration. In common with other construction projects, tunnelling will encounter legislation concerned with authorisation, through Parliamentary procedures for major work, or local planning consents by means of application and consent. In Britain, these procedures often involve an Inquiry without imposed timetable. Formal submissions may be required for Environmental Impact Assessment (EIA) and for other consequences of the proposal. Particular rules will normally apply in relation to the several forms of infrastructure, with certain powers already granted to statutory providers of services. Where innovative arrangements between the Parties to a Project are contemplated, departing from traditional relationships, these clearly need to be set on a sound basis in relation to the law of the country. The Public Finance Initiative (PFI) and any of its derivatives in terms of management of services for the public by private organisations such as Build-Operate-Transfer (BOT) fall into such a category. Additionally, dispute resolution requires appropriate provision (Section 8.5).

Underground work is demanding on specialist skills and in their integration. Whatever may be the legal requirement for advertising projects for prospective tenderers, there must be safeguards to ensure that only those with full competence and with a professional attitude appropriate to the particular project are invited to compete or participate.

For underground works, there are additional issues for possible legal concerns. Many centuries ago, the phrase was coined: 'cujus est solum ejus est usque ad coelum et ad infernos' i.e. the ownership of the surface extends to the sky and to the depths. Thus, while ownership upwards has been clipped to permit free air navigation, land tenure rights below the ground may be more murky; they may or may not include mineral rights and usually exclude the right to extract water. Furthermore, there will be duties to avoid causing damage to others by work undertaken below ground. Different countries have enacted different laws concerning the right to make use of the subsurface. A sensible present view is that the landowner has only an interest in that part of the sub-surface that he might reasonably develop beneficially.

An international study undertaken by the International Tunnelling Association in 1990 (Sterling 1990) summarised the different regulations of 19 participating countries relating to the development of the sub-surface for different applications. Generally it was found that such development was the sole prerogative of the landowner, subject to specific conditions and to conformity with planning regulations. In general, zoning of use of the surface did not confine the use of the sub-surface except at points of surface works for access or other related purpose.

The earliest Metros were sited beneath public roads to avoid the need for obtaining special easements. When this was no longer practicable, the right to construct underground infrastructure services was generally acquired by negotiation or through statute, with provision for compensation for direct loss of value or for damage, even occasionally for nominal loss of value.

In view of the increasing perception of the planned use of the sub-surface, steps to avoid negative activities of local conflict such as the obstruction to a future Metro by the construction of deep foundations or, positively, to encourage multiple combined use of a tunnel, have been advocated. Many countries now find benefit in the more coordinated use of the sub-surface with ability for expropriation by means of legislation. The ITA statement on this issue (Sterling 1990) reads:

'The sub-surface is a resource for future development similar to surface land or recoverable minerals. Once an underground opening is created, the sub-surface can never be restored to its original condition and the presence of this opening can affect all future uses of the surface and the subsurface in its vicinity. These factors require responsible planning for all uses of the underground to ensure that the resource is not damaged or usurped by uncoordinated first uses.

The awareness of the underground option among planners, developers and financiers should be increased so that sub-surface planning issues are properly addressed. Sub-surface planning should be an integral part of the normal land use planning process.

National, regional and local policies should be prepared to provide guidelines, criteria and classifications for assessing appropriate uses of underground space, identifying geologic conditions, defining priority uses and resolving potential utilization conflicts. Site reservation policies should be established for important future uses and for especially favourable geologic conditions.

It is recommended that every region or city establish a permanent record-keeping system for the maintenance of detailed records of the use of the sub-surface. This record-keeping should be coordinated by a single agency to ensure compatible and complete records and should include "as built" records rather than project plans. Records should include activities, such as ground-water extraction and deep pile foundations, which affect the potential use of the sub-surface but which may not be classified as specific sub-surface facilities.'

In Minnesota and Chicago, cities with particularly favourable geology, enabling legislation has been enacted to remove what were seen as institutional barriers to future desirable sub-surface development. These powers not only provide for acquisition of the underground space, with funding and construction of development, but also the protection of the sub-surface against damage which might inhibit subsequent planned development. In Kansas City, where limestone mining by pillar-and-stall has provided extensive space for commerce, offices, laboratories and storage, building regulations have also been revised to take account of the special conditions pertaining to activities undertaken below ground.

Apart from questions of damage caused by subsidence, matters of less direct consequence need to be addressed, e.g. changes in water-table affecting adjacent property, as occurred when dewatering for Amsterdam's Metro in the 1960s affected timber piling to buildings in the vicinity. Another more bizarre instance is recorded [Ownership of subterranean space, W.A. Thomas, in Sterling (1990)] where natural gas stored by a company in caverns escaped to a gas reservoir belonging to an adjacent landowner who sold it. The company sued but the court dismissed the case, holding that the escape of the gas constituted a trespass.

3.4 Competence in planning

During the planning process, issues which involve the relevant Parties in the optimisation of the project will be encountered, particularly concerning standards of safety, operation and maintenance, and these should be identified and recorded, for resolution at an opportune time. For the 'unfamiliar client' in particular (Section 3.1.2) heed needs to be given to the notice of policy decisions by certain times to avoid delay in the development of planning.

A document prepared for the Permanent International Association of Road Congresses (PIARC) (Muir Wood 1995) provides advice to 'new clients' for road tunnels, much of which has more general application to 'new clients' for many forms of construction projects, emphasising the nature of decisions to be made during planning, which will be affected by intentions in operation, to allow optimised total life costs to be prepared. For those road tunnels in particular where demand may be expected to increase markedly with time, there may be benefit in a low-cost initial construction designed to permit subsequent fitting of equipment and services to enhance operational standards without the need for major reconstruction or interruption of service. The PIARC Report referred to above (Muir Wood 1995) states that:

> 'The design of a road tunnel requires the effective amalgam of expertise between those familiar with transport planning, with construction and with successful operation of such projects. In the absence of existing experience of a road tunnel authority, it will be necessary to introduce a "surrogate operator" into the planning team, an engineer with appropriate experience who is given powers to ensure that the project planning is undertaken in a rational and coordinated manner.'

The need for efficient communication between members of the planning team is emphasised and discussed in Chapter 2.

Specific to a road tunnel are the questions of the degree of autonomy of the project, the financing base and the raising of tolls which may indirectly affect many other aspects of operation. Other particular issues include the following:

1. *Composition of traffic.* Safety in service depends on adequacy of provision for future demand, in terms of tunnel cross-section, gradient and

curvature. There is need for statistical predictive information on traffic. The tunnel should generally respect the standards of adjacent roads (checking the ratio of user benefits to tunnel costs); where this is impractical, variation in standard should be made well clear of the tunnel portals.

2. *Provisions for accidents.* For short tunnels (say, < 1 km), any specific provision depends on importance of the route. For medium length tunnels (say, 1–3 km), a hard shoulder should be considered to allow traffic to pass a stalled vehicle at low speed. For long tunnels, emergency stopping lanes at intervals along the tunnel is an alternative arrangement, combined with provisions for escape. In all instances, the optimal means of respecting the safety requirements will be project-specific, recognising the very different incidence of additional cost for particular facilities requiring tunnel enlargements in relation to different forms of tunnel construction. Safety requires a holistic approach.

3. *Provisions for equipment.* Equipment for normal operation such as lighting, ventilation and traffic control should be supplemented to deal with emergencies of fire and collision.

4. *Control and monitoring facilities.* Such provisions depend on the nature of the traffic and the degree of autonomy in the operation of the project.

5. *Junctions.* Any feature giving rise to weaving of traffic within the tunnel should be avoided, thus junctions should be set well clear of tunnel portals.

6. *Provisions for maintenance in service.* If maintenance during operational service is intended, provision needs to be made for access for operatives, plant and materials. Ventilation will need to be adjustable to meet standards appropriate for long-term exposure to tunnel fumes.

7. *Drainage.* Drainage needs to include interceptor chambers and traps to avoid risk of the spread of fire by inflammable liquid. Tunnel gradients should respect the needs of gravity drainage.

8. *Risk.* The incidence of accidents in tunnels does not exceed those on the open road but the consequences, and the problems of dealing with accidents, may be more severe. Risk assessment should consider the possible consequences of accidents, particularly of fire or explosion, and their mitigation. There is also need to consider effects on the tunnel structure and, most importantly, on the continued operation of vital tunnel services and communications following an accident. A rigorous consideration of possible consequences may follow the procedure of constructing an 'event tree' (Blockley 1992). Limits may be imposed on traffic allowed to use the tunnel, in respect of risk of explosion or fire. If certain categories of vehicle are to be excluded,

it may be expedient to make differential risk assessments in relation to alternative routes, also to the practicability of enforcement of rules of exclusion.

Comparable issues affect tunnels for other purposes.

3.5 Coordinated planning of projects

3.5.1 Multiple-purpose projects

Bridges readily serve for more than one purpose, e.g. as road and rail combined or as road bridges carrying service pipes and cables. Tunnels are less readily adaptable since the problems underground will include the provision of access for inspection and repair, segregation, protection in the case of accident or fire in the tunnel or as a result of damage to the service pipe or cable (e.g. escape of gas or water). When a major barrier is pierced by a tunnel for the first time, multiple usage merits consideration, especially for 'inert' services such as optical cables. Multiple service ducts are widely used at shallow depth in cities, the absence of common terminal points often militating against such ducts being constructed at depth. Immersed tunnels represent a partial exception to the rule, since single tunnel units providing for multiple and separated ducted spaces for road, rail and for services will cost less than separate units for each purpose. In locating and protecting services, the risk and consequences of damage by accident or fire needs proper respect. (Also see Section 3.4, point 8).

3.5.2 Serial planning of projects

Where, as is usual for tunnelling, the horizon for planning is at a distance beyond that adopted for identifying the least cost option by the accountant, merit may well be found in choosing an option which is a good fit with longer-term development, e.g. by oversizing a stage of drainage tunnel or routeing a water tunnel to suit a subsequent stage of extraction or supply.

As an early example of more ambitious serial planning of projects, Allport and Von Einsiedel (1986) describe a solution to the practical problems in planning infrastructure projects in the Philippines. This stratagem has more general application elsewhere in circumstances in which several aspects of infrastructural development are competing for funding, particularly in developing countries where demand for potentially viable projects exceeds availability of funding.

In the past, these authors state that they found that individual projects had been allowed to proceed without a common basis for evaluation, in a climate in which optimistic forecasts for economic growth had encouraged lax discipline in control. Changing money values, affected by inflation and devaluation, with consequences to contractors' profitability, added to problems in forecasting disbursement needs. Thus, even those projects authorised to proceed encountered problems in funding, causing long over-runs. Many of the departmental agencies established to develop programmes were following independent paths, forming new agencies with ill-defined mandates. In such circumstances, while government made strenuous efforts towards progress, the planning process itself was found to be ineffectual.

One of the contributions to solving the infrastructure planning problems in Manila, described by Allport and Von Einsiedel was the Capital Investment Folio (CIF) process, planned to complement rather than supersede existing systems for the allocation of resources. An essential element of CIF concerned the establishment of investment priorities across the public sector. The first problem to be solved was the fact that national funding was by sector, e.g. transport, water, rather than geographical area.

The practical solution required agreement with a new body comprising city and national agencies, the Inter-Agency Technical Working Group (IATWG). Initial cooperation of the existing Agencies was mixed, but in due course a list of projects was prepared across the several sectors of infrastructure. Next, sets of sectoral strategies were examined which allowed development of a coherent overall strategy within a sector. Then, interaction between the IATWG and the Agencies allowed consideration of alternative allocations of incremental capital investment. This stage also allowed development of linkages between the proposals of associated Agencies, e.g. water supply, flood control, sewerage, waste disposal. Projects which might attract private investment were also separately identified.

Where projects were already under construction, assessments were based on the cost assumptions for completion and the consequential benefits. For new projects, tests were made against several scenarios, favouring those which indicated robust benefits and which were flexible in adaptation to the uncertain unfolding scene in contrast to previous rigid planning systems. Figure 3.7 (based on Allport and Von Einsiedel 1986) indicates the screening and evaluation stages in the establishment of an overall Core Investment Program (CIP).

Consideration was also given to other criteria including the matching of projects to the capabilities of the Institutions responsible for their direction, and the affordability of the elements of cost falling upon Local Authorities. The process was understood to entail gradual evolution, gaining experience, towards an improved allocation of resources, particularly in relating investment priorities and timing of projects between the several sectors.

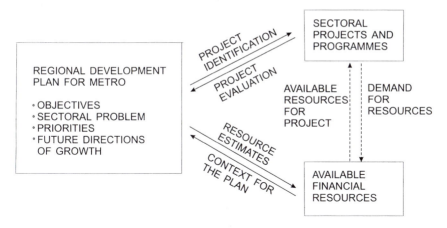

Figure 3.7 CIF planning process and interlinkages (after Allport and Von Einsiedel 1986).

3.6 Issues of procurement of concern to planning

Procurement and associated contractual issues are discussed at some length in Chapter 7. During the initial planning stages, particular thought needs to be given to such contractual aspects as:

1. Provision of continuity in conceptual planning if the client is to change or if the intention is to pass responsibility for design to another Party (BOT, for example). There is then a likely need for a degree of 'nova-tion' whereby the obligations of a designer of the concept or of particular features are taken into the new Party. A break in concep-tual thinking at this stage may well lead to inability to derive benefit from an innovative approach that needs continuity of development into project design and design of the construction processes.
2. The need for cost estimates to take account of the contractual arrange-ments. If a partnering concept is intended, optimal means for dealing with risk may also be devised, with optimal consequences for the control of costs. If, at the other extreme, construction risk, including geological risk, is to be passed to the Contractor, the costs must be calculated against the most unfavourable foreseeable scenario of all risks eventuating, with a margin for those unforeseeable and for possible litigation in circumstances which may be worse yet – and generally for tunnelling such costs may be excessive. (Figure 3.5). In the absence of equitable risk sharing, the financial benefits of the *design* process as described in Chapter 2 cannot be realised, hence the need for additional allowance for increased cost.

3. Where legislation or custom requires that some form of open tendering
    is to be operated, where Parties are to be selected on least cost without
    regard for special competence, additional provisions should be made
    to cover risk overall. The consequences are liable to be so unfavourable
    and unpredictable that strong advice should be given to make an
    exception to such a ruling on the grounds that it is totally unsuited
    for work so dependent on experience, expertise and quality as
    tunnelling.

3.7 Reliability of forecasting

Generally, experience is not good in the reliability of estimating for major
infrastructural projects, particularly where financial viability requires estimates of revenue. There is merit in identifying what are the reasons for
this relative lack of success, what are the specific features of underground
projects in such respects and what may be done towards improvement
for the future.

For sub-surface projects, the inherent and highly site-specific nature of
the ground adds one more uncertainty to the problems of forecasting into
a future scene of unknowable political and economic circumstances. By
first identifying the main areas of uncertainty and then addressing how
these may be limited or specifically qualified in estimates, policies towards
overall risk control may be more effective. The main areas of uncertainty
are described below.

3.7.1 Political influence

Where projects have strong political concerns, there are suspicions,
supported by evidence, that figures adopted by Governments, and possibly
others, tend to involve manipulations or concealments of particular factors,
in order to appear to support a particular objective. Skamris and Flyvbjerg
(1996) provide evidence of suspicions of such activity for Danish transport projects, and there is no reason to believe that Denmark is more
prone to economies in the truth of such a nature than other nations. In
the many years of preparation for the Channel Tunnel, Government
Reports provide several examples of:

*   direct comparison between options of quite incomparable reliability
    of basis;
*   the application of contingency sums of an arbitrary nature;
*   the acceptance of revenue figures extrapolated well beyond any rational
    basis for projection.

3.7.2 Authorship of estimates

There are several possible causes for distortion in early estimates of projects with contributions under these categories:

1. Early estimates of cost may be sought from management consultants or economic advisors who have little understanding of technical issues, particularly those specific to tunnelling, and frequently even less of their own deficiencies in such respects. Preliminary estimates are therefore based on historical costs of comparable, but by no means similar, projects without appreciation of the great spread of costs covering the most favourable and most unfavourable circumstances.

2. There are examples of estimates of cost being prepared by those without great scruples for reliability but who wish to remain associated with what may appear for them (at least until the date of abandonment) to be a profitable assignment. Such estimates will be well below a viable figure. One, fortunately rare, variant of this practice is that of the Engineer (acting in this contractual capacity) who initially undertakes to his client to fulfil a project within a certain sum and then encounters embarrassment between Client and Contractor when unforeseen problems emerge which would cause the figure to be exceeded.

3. The recent practice of fragmentation of responsibilities for the elements of project development has resulted in advice on different aspects of a project being commissioned in separate packages from different sources. Apart from the consequential inability to introduce optimisation, this practice will also tend to stifle innovation and to introduce by one fragment of the project sources of risk to the success of another without realisation by either, since they have operated at arms length from each other (see Section 2.5). Innovative ideas, of potential value, would inevitably need to be tested and evaluated, across all relevant aspects, a procedure excluded by the process of 'design-by-fragmentation'. An engineer, engaged for a confined element or phase of a project and hence knowingly deprived of the opportunity to develop new ideas, would be naturally reluctant to discuss the possibilities (there being moreover examples of inferior unsuccessful attempts at implementing the good ideas of others where the original authors – in a litigious age – have been subsequently blamed for the failure).

4. Some economists have tended to extrapolate trends of demand well beyond credible dates for acceptable percentage increase, having regard to associated developments that would be needed elsewhere to permit such general increase in demand. On the other hand, departments of central and local government tend to depress estimates of figures

of demand in order to minimise capital expenditure in the short term, without regard to the often massive increase in costs to satisfy demand by piecemeal construction.

3.7.3 Economic and political factors

Assumptions adopted in estimates need to be stated with clarity, with evidence for calculations based on correlations between economic indicators and costs (and revenues) for the project. The question needs to be asked as to the risk of new legislation affecting the viability of the project.

3.7.4 Timing of completion

Estimates of cost are normally related to a particular date, indexed backwards and forwards. There may yet remain other factors which depend on particular timing of the project. For example, viability may be affected by relative timing with associated projects, projects may be timed to be available for particular events (e.g. World Olympics) which may not only help to raise first year's revenue but also to establish habits of usage.

3.7.5 Development of competitors

Where there are competitor projects, technological development and marketing strategy, the 'game plan' needs to be considered for each. For example, estimates for the revenue earning capacity of the Channel Tunnel needed always to be established against competition by the ferries of the future, not confined to those operating at the time of the estimate.

3.7.6 Ranges and qualifications

Too often the estimator is required to produce a single figure where uncertainty exists. What should this figure be: the worst extreme, the most probable, one which makes reasonable provision for uncertainty? All figures should be qualified to discourage their misuse. It is impossible to prevent their use out of context but this practice should then be apparent to an auditor.

3.7.7 Attention to 'climate of risk'

Pugsley (1966) has drawn attention to the nature of technical factors, such as simultaneous innovations, which may, in combination, give rise to high risk. This concept may be extended into novel areas including investment, untried contractual provisions and the risk of political interference.

3.7.8 Changes in requirement, including uncertainty and vacillations

Perhaps the commonest single cause for high increases in cost of an underground project is the late requirement for change, often as a result of too perfunctory a planning stage. A great proportion of the expected reduction in uncertainty of cost during the planning stage (Figure 3.5) arises from firm decisions affecting requirements. Project financiers are often too concerned in seeing an early return from their investment to appreciate the merits of deliberation in the early phases of project planning when the foundations should be laid for sound project definition. Causes of late change may be internal to the project or external. A prominent external cause relates to upgrading of the requirements for health and safety. A thorough risk analysis during planning may help to avert problems caused by subsequent hiatus when standards are reviewed too late. For the Channel Tunnel, the effect of numerous changes to requirement were exacerbated by the partial opacity of the Contract Documents. For this project, the Inter-Governmental Commission, responsible for setting safety standards through their Safety Commission, established only after construction had begun, made many late requirements for upgrading safety, while construction, including work commissioned from the main sub-contractors, was already in progress. Directly or indirectly such changes made considerable contribution to the 80% rise in estimated cost of the project. Projects of such size and magnitude require a risk-based approach to planning and design (Section 2.1.3), which should eliminate or greatly reduce the impact of high elements of uncertainty before these can be the cause of such major distortions of cost. Questions of risk may extend well beyond the project itself, including factors material to success and the actions of other parties whose cooperation is essential to achieve the objectives of the project. Once again it needs emphasis that not only may questions of changes in design be involved but also delays on account of the length of lines of communication and the complexity of particular issues once the matters external to the project have been adequately resolved. Planning needs to foresee and control such issues.

3.7.9 Contractual relationships

As described in greater detail in Chapter 7, the contractual relationships have a great impact on the development of estimates as the project proceeds through its several stages (Figure 3.5) There are three inter-related issues:

1. Where risk is deliberately (or by oversight) loaded on to the contractor, tenderers are likely to take a highly pessimistic attitude as to the eventuation of risk; thus the tender price will be expected to be high, with

no prospect of reduction when circumstances are found to be more favourable. On the contrary, should the contractor have underestimated, every opportunity will be grasped for establishing a cause for being misled by information provided by the employer. This is by definition a 'brittle' contract, susceptible to litigation.

2. The more the risk is placed on the contractor, the less the prospect of a concerted risk assessment which, to minimise risk, will almost certainly require flexibility by all parties who are in a position to contribute to measures to reduce risk, as described in Chapter 2.

3. In the absence of equitable sharing of risk, the relationship between the parties, at least at contractual level, will tend to be poor; problems which otherwise, by wise anticipation, might be circumvented or readily controlled by a cooperative approach, will tend to fester. Each side will, perforce, become more concerned, by the nature of the contract, to establish the fault of the other than to find an economic solution.

The proclivity of lawyers to inject confrontational attitudes into contract relationships and more ubiquitously should be strongly resisted and clients should appreciate the likely cost of bad advice in such directions. Unfortunately, too often decisions of such elementary error are made by those who do not understand the likely seriousness of the consequences, at so early a stage in defining the project that no other voice is powerful enough to counsel a wiser course.

3.7.10 Tendering processes

Tendering processes may also have the effect of preventing the favourable outcome of a project. This feature is further developed in Chapter 7. Here it is only necessary to state that any practice which neglects questions of skill, technical competence and availability of adequate resources in appointing a tunnelling contractor is likely by sowing the wind to reap a whirlwind. Where the engineering expertise to design and possibly to supervise construction is appointed by competition on cost, a fuse is ignited towards a more explosive disaster.

3.7.11 Inflexible programming

Underground projects often entail interactions between separate operations or separate contracts. These may entail 'interleaving' or operations being undertaken end-on. Critical dates need to be chosen with great caution. Programme 'float' may float away on account of some new constraint which, even without overall loss of time to any contract, may lead to reversal of the order of undertaking specific operations. Delays caused by inter-contract conflict tend to be expensive; even more so may be the implementation of stratagems designed to make good loss of time. It is, of course, right to have optimistic targets for progress which will reward efforts for continuous improvement, but these should always be accompanied by realistic contingencies to avoid conflict. Better by far is the objective of maintaining flexibility between the several parties in recognition that, without compromising obligations, adequate understanding of the essential features of each element is a *sine qua non* towards finding the optimal solution.

3.8 Practical examples of success and failure in planning

Many of the problems encountered in tunnelling, and many of the causes for the elimination of the tunnelling option, have arisen during the planning – or non-planning (i.e. where planning has overlooked the option of tunnelling) – phase. A few examples are described below and many more of the examples of Chapters 8 and 9 stem from basic planning errors.

A surface plan for bypassing road traffic to the South of London, termed the Southern Box Route, was developed by the Greater London Council during the early 1970s. This scheme encountered much opposition on account of its land-take, effect on the environment and severance of communities. A report was subsequently commissioned (Greater London Council 1973) to review the opportunities for constructing roads in tunnels under London in general and in replacing two sections of the Southern Box Route in particular. This Report demonstrated that, whereas for one section the tunnelled solution would cost about 80% more but would take less than 45% of the area of land, for a second section where tunnelling conditions were more favourable, the tunnelled solution would cost very little more than the surface scheme and would only sterilise about 23% of the area of land. One main conclusion of the study was that if a tunnelled solution had been considered from the outset a route might have been selected to take better advantage of the topography to the South of London in such a fashion as greatly to simplify access roads connecting to surface routes and to offer greater benefits against a surface option. In the event, the Southern Box Route project was abandoned.

For the City of Bath, a project to relieve road traffic by means of tunnels was thwarted by dysfunction between the planning and the engineering processes. Planning commissioned by the City Architect and Planner concluded that a solution might be designed around a short in-city tunnel combined with a by-pass tunnel. This conclusion was based upon assumptions of tunnel unit costs for the two elements which were approximately 50% and 200% of reasonable estimates respectively. One consequence of this error was to bring the approaches to the by-pass tunnel, in view of the over-estimate of its cost, excessively close into the centre of the city which was at that time already suffering the results of redevelopment too close, in the view of many, to the core of the Regency City of Bath. The project was abandoned and the 'window of opportunity' in consequence passed by. The traffic problems continue to increase year on year, only partially relieved by a surface bypass route using mainly existing roads to the east of the City.

A Parliamentary Bill for the construction of the twin-bore, two-lane Clyde road tunnel had been obtained in 1948 prior to any serious consideration of the problems of construction. Apart from tortuous initial approaches to connect the tunnel to existing roads along each bank of the River Clyde in Glasgow, the curious W profile of the road surface in the tunnel represents a not altogether successful endeavour to contain two lanes of traffic to full height gauge within the inadequate tunnel diameter described in the Parliamentary Bill (Morgan *et al.* 1965).

Second stage site investigations during the early stages of planning for the Ahmed Hamdi road tunnel beneath the Suez canal disclosed that the original crossing would have entailed tunnelling through water-bearing sandstones of low strength. The location of a major fault allowed the tunnel to be resited several kilometres further to the north in mudrocks, considerably more favourable for tunnelling. Advantage was taken of this benefit in a somewhat comparable manner, in tilting of the ring of the 27 km approximately circular CERN collider path, near Geneva, and thus allowing the tunnel to be sited predominantly in the more favourable weak sandstone 'molasse' and not the underlying water-bearing limestone.

Planning of the work of providing a new secondary lining to Brunel's Thames Tunnel proceeded on the basis of least cost without regard to the exceptional merit of the original project, in the absence of a formal registration of the tunnel for conservation (Roach 1998). An increase in cost of 250% (£6.3 M to £23.2 M) was attributed to this lack of foresight (Section 1.4). The requirements for engineering conservation, if considered from the outset, would not have caused any increase in cost. As constructed, respecting engineering conservation, Brunel's lining may be expected to remain intact and thereby ease subsequent problems of repair of the internal concrete lining.

A project of irrigation and drainage to rehabilitate a massive farming project to the West of the River Nile would optimally feature a drainage tunnel to the Mediterranean through low limestone hills, constructed by a technique of Informal Support (Chapter 5). An essential feature would be continuity of the engineering design of the project through the construction phase. The alternative would be based on a considerably more expensive segmentally lined tunnel. The Client Authority was unable to accept the condition of continuity and the tunnel option had therefore to be abandoned for a less satisfactory surface canal around the range of hills.

Chapter 4

Studies and Investigations

> *If you do not know what you should be looking for in a site investi-*
> *gation, you are not likely to find much of value.*
>
> 1968 Rankine Lecture, Rudolph Glossop.

4.1 The methodical acquisition of data

Studies for a tunnelling project will be required on a number of different
aspects affecting construction and operation. The operational requirements
(Chapter 2) will determine the aspects in which operational studies will
be needed. For example, operational studies will provide estimates of
demand for the project, criteria to establish viability in terms of cost and
revenue, social benefits and other aspects which may be selected to be
tested against the several options – including that of 'do nothing'.

For construction, studies will be required of all the principal features
affecting definition of the optimal project, related to the cost and time
for its execution. These studies will concern initially the interpretation of
the geology as to its engineering consequences.

4.1.1 Studies relating to operation

Studies concerning the functioning of the proposed project must be
specific to the project, to the overall purpose and to the stated objectives.
Tunnelling may well form only part of a larger scheme. To permit initial
planning to be undertaken in good time, as emphasised by Chapter 3, the
initial studies relating to the underground element may need to be undertaken a considerable period ahead of embarking on more detailed planning,
to allow adequate flexibility in the planning of the scheme overall.

Studies on project-specific operational policy may be described under
these categories:

1. *Demand*, acquiring data on which time-dependent demand estimates
   may be based, against a range of scenarios, expressed sufficiently

explicitly to facilitate future modification of predictions as assumptions may change and when trends with time become more evident. Where appropriate, demand will be based on charges for usage in relation to competitor projects.

2. *Financing options* for projects which may draw upon a measure of private finance, in equity and borrowings, or for public funding from different sources.

3. *Quality standards* in relation to performance, safety, project life and other factors affecting overall utility.

It is to be noted that all such features, vital for a successful project, will depend upon 'soft data', i.e. upon informed opinion and not upon ascertainable fact. It is essential that the manner in which studies have been undertaken be explicit and thoroughly documented, particularly in view of the long time-scales which may be associated with the several stages in which the studies may need to be undertaken.

Studies of this nature are common to all construction projects and have no particular relationship to tunnelling except in so far as tunnels present special difficulties in modification, as demands or standards may change, subsequent to initial construction. There are however specific features of tunnels which merit emphasis in relation to financing and legal issues:

1. A tunnel project will normally need to be preceded by application to Parliament or a comparable body for formal sanction. This will be a protracted process and will in consequence occur early in the period of project planning. It is necessary to ensure that the application confers adequate flexibility to suit the adjustments that may be desirable as a result of subsequent studies in concept, overall dimensions and in siting, or may be necessary from considerations of safety.

2. A tunnel provides limited scope for incremental development. In consequence, until completed it has little intrinsic value. Studies in means for financing must therefore take account of the different degrees of exposure to risk as the project develops, particularly in relation to risk pre- and post-completion.

The viability of a privately funded traffic tunnel will be based on estimates of demand in relation to charges in the form of tolls or credit transfer (as part of a Private Finance initiative) for vehicle transit. The higher the charge in relation to competing links or modes, the lower the demand, as illustrated by Figure 4.1. In consequence, gross revenue will take the form of an inverted-U curve, being zero for zero unit charge and approaching towards zero when the unit charge becomes excessive. The operating costs will have a fixed element, related to servicing capital and undertaking essential maintenance, and a variable element dependent on

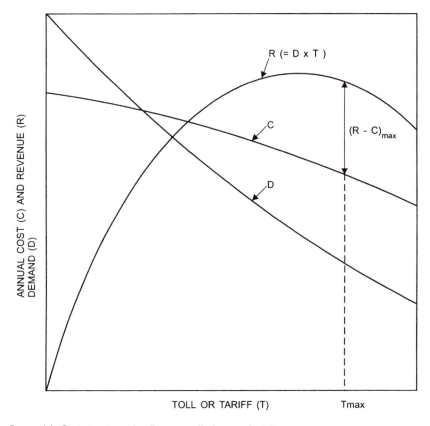

Figure 4.1 Optimisation of tolls or tariffs for profitability.

demand. As illustrated by Figure 4.1 it is possible in consequence, on the basis of marketing studies and estimates of costs, to find the approximate level of unit charge to maximise net revenue. Other factors, such as reduction in traffic congestion elsewhere on the network, may introduce social factors to influence whether the unit charge should be pitched somewhat above or below that predicted as commercially optimal, with possible cross-subsidy for a privately funded project.

4.1.2 Studies relating to the execution of the project

Apart from investigation of the ground, feasibility of construction will be concerned with such questions as:

- access and locations of working sites;
- availability of resources, including particular skills;

- development of plant and techniques best suited to the project;
- environmental and social issues;
- interests of other users of the ground surface and sub-surface;
- restrictions on working, e.g. noise and vibration limits, height restrictions near airports;
- spoil disposal;
- recent experience in comparable projects elsewhere.

All features concerning tunnel construction, such as access for the work force, for plant and materials, needs for working sites, for electric power and other services will be related to the preferred scheme of construction. Studies of such needs will therefore become focused as the scheme of tunnelling becomes defined. At each stage of project definition, the potential adequacy of such features needs to be considered. For example, the power requirement for a tunnel driven by TBM with electrically powered haulage may be in excess of reliable available public supplies and require in consequence a project power-plant.

Studies appropriate to a particular tunnel may relate to many aspects. For example, basic data affecting the logistics of a tunnel project in an isolated area will be necessary in order to establish potential problems and to provide a data base for estimating costs. Such data should be collected and presented in a growing statement of factual project data for general use of those involved with the project.

Every tunnel will require the acquisition and interpretation of information relating to the ground and to the prediction of its behaviour as a consequence of tunnelling. Studies of research and of practical experience related to other projects may be valuable in selecting features of the most appropriate scheme to satisfy the demand. Thus, the major element of studies to complement planning and design will concern the nature of the ground and how this inter-relates with possible schemes of construction (Figure 3.3). As a matter of definition, the term 'site investigation' (s.i.) follows the example of others (e.g. West *et al.* 1981) in embracing the total process of determining the nature of the ground in respects relevant to the options for construction, while 'ground investigation' is defined as that part of 'site investigation' which entails geological and geophysical investigations commissioned expressly for the project. The progress of site investigation will generally follow the sequence set out in Table 4.1. Stage I will be essentially a desk study, Stage II entails predominantly work in the field and the laboratory, together with interpretation of the acquired data from Stages I and II, while Stage III will be based on observations during construction together with further development of interpretation of all available data.

A practice has developed in Britain (not in Northern Ireland where borehole records are in the public domain) whereby those who have

Table 4.1 Sequence of site investigation

Stage I: the desk study which assembles from accessible sources (if any) geological maps, associated memoirs and monographs on specific relevant aspects of particular interest. Other local data may be available from private sources or from the national Geological Survey or Office. Where no detailed information exists, this stage will start by assembling data of a more general nature with recognition that the subsequent stages start from a relatively low base of local knowledge and hence of wide generality of information required.

Stage II: the pre-construction site investigation, which may be conducted in more than one phase, providing information of value to the project, generally in increasing degree of detail and more focused towards the preferred tunnel route and options for construction. A final phase of this stage may entail trial shafts or tunnels which may be visited by tenderers.

Stage III: further investigation undertaken as part of the project itself.

commissioned s.i. for construction projects establish 'commercial – in confidence' rights to the results. While this may appear to have a valid short-term commercial justification, it is contrary to the public interest, will tend to lead to predictions about the ground being based on suboptimal information – even by those who may be holding one part of the data – and could well contribute to an underground accident which might have been prevented by open access to such 'commercial' records held by others. The British Health and Safety Executive might well reflect upon the desirability of abandoning such a practice, prior to a consequential accident.

Much has been written on the subject of relating geology to specific engineering behaviour of the ground with the accompanying engineering problems and opportunities, the best of these by those who have practical experience of relating geology to the *design* process. See, for example, Legget (1979), Legget and Hatheway (1988) and, for the special case of a single major project, *Engineering Geology of the Channel Tunnel* (Harris *et al.* 1996).

Site investigation should be seen as an integral part of the *design* process, of yet more vital importance to the design of the construction process, including the design of the means of construction, as to the design of the permanent works for the tunnel. It is only by such recognition that the appropriate elements of site investigation may be designed, how much determined, to what purposes and with what specific objective indicators of achievement. Adequacy and quality of s.i. represent an indispensable key to success of the project overall. This is the primary justification for the s.i. to be designed and managed by those who will be responsible for its application to the overall scheme of construction and, as a corollary, by those who understand the dependence of the construction processes on adequacy and reliability of specific information about the properties of the ground. At all times and for all stages of a site investigation there must be particular objectives, particular expectations and a keen understanding of the impact of the findings in relation to the tunnelling options. Where a departure from expectation occurs, its significance needs to be promptly assessed. Does this entail rethinking of the s.i. strategy or a shift in emphasis? An anomalous result should not be dismissed as a 'rogue' without good reason; it may represent the most important feature of the s.i. In other words, progress of the s.i. affects not only the other aspects of planning and design but also possible changes of the s.i. itself.

Features of the several stages of s.i. are described in succeeding paragraphs. As described in Chapter 5, site investigation needs to develop in association, by planned interactions and frequently in an iterative manner, with the planning and conceptual design of the project: what it is to comprise and how it is to be achieved. Thus, s.i. is vital to the preparation of tunnelling options while the definition of these same options will contribute to the design of the s.i.

The starting point for ground investigation must depend on the nature and extent of pre-existing knowledge of the ground in the locality. An undeveloped part of a country at an early stage of geological mapping provides a very different starting point from a city much perforated by tunnels at depths comparable to those of the proposed project. For the former, a walk-over survey combined with the study of topographical maps and aerial photographs will be expected to provide first evidence of surface features and lineaments indicating faulting. Good Codes of Practice exist on the general aspects of site investigation (e.g. BS 5930 1981) and what follows has specific relevance to tunnelling.

Boreholes will form a central part of the investigation, their nature, number and siting depending on the information required. The questions to be addressed include:

- To what extent is geological history, structure and stratigraphy understood?
- Will the form of tunnelling require specific information on particular features of the ground types? If so, what specific data are required?
- How does the variability of the ground affect the extent of detail required by the investigation?
- Is the presence of specific faults, intrusions, unconformities or other anomalous features suspected? If so, how should this affect the design of the investigation?
- Is water a potential problem? If so, what is required to define the nature of the problem and potential solutions in relation to expected forms of construction?

- Are any hazards expected, such as the presence of methane, ground contamination or corrosive ground-water?
- Does the nature of the ground present special problems of susceptibility to damage of structures or services in the vicinity?

Boreholes for tunnelling have been used predominantly as the source of descriptive logs, cores and samples for inspection and testing, more recently for increased extent of *in situ* testing (e.g. penetrometers and vane testing for weak soils, pressuremeters for stronger soils and weak rocks). Increasing use has also been made of 'down-the-hole' testing derived from the oil prospecting industry, for determining the physical characteristics of rocks and of discontinuities, also of water flows. For example, for the Channel Tunnel, since 1964 sonic and electrical logs were used to identify particular geological horizons of the chalk (Muir Wood and Casté 1970). The exploratory holes for the Rock Characterisation Facility for nuclear waste disposal at Sellafield (UK Nirex 1993) represented the state-of-the-art of the time, providing geophysical logs representing in great detail the projected surface of the borehole and of water inflow, at a considerable cost (~£10M for each 1000 m borehole)

Inclined boreholes are expensive but may provide the only practical means for investigating otherwise inaccessible features. For example, the Piora Basin (a 'sugary' dolomite bearing water up to 10 MPa pressure) has been investigated in this manner (Hackel 1997) by a deflection borehole from an exploratory tunnel for the St Gotthard Base Tunnel (Flury and Rehbock-Sandes 1998). Steered boreholes from the shore may also be used to explore the ground under-water without incurring the cost of drilling from barge, jetty or platform.

Tunnels will, where the choice exists, be sited in ground favourable for the selected form of tunnelling, using equipment designed to cope with expected problems. Working shafts, and shafts for other purposes, may need to be sunk through a variety of types of ground. In consequence, the scheme of site investigation should include boreholes sited in the close proximity of prospective shafts, to include investigation of the problems specific to the shaft-sinking including possible effects of ground loss or water inflow. The pattern of boreholes should respect geological features and their variability, for urban tunnels relying on the availability of drilling sites, removing unacceptable uncertainty on aspects of the ground which could have significant effect on the scheme of tunnelling. Boreholes should avoid intersection with tunnels and should be effectively back-filled with a bentonite/cement or similar material to provide stability and prevent passage of water across aquicludes.

The levels and mineral content of water encountered by the borehole may provide important data, including seasonal or other variation. Study of water flow patterns may justify pumping tests, each using several observation wells so that features of ground-water flow may be based on the shape of the cone of depression, and possibly of detail of its recovery when pumping ceases. A more popular, but less readily interpretable, method is that of pumping-in tests between packers in a borehole. The basis of analysis of such a test is described in Appendix 5G. Here it is pertinent only to remark that the distance between packers may be important where permeability variation is to be calculated for rock, also that such packer tests measure predominantly permeability transverse to the line of the borehole.

There is considerable choice in the type, quality and, in consequence, cost of different forms of drilling, and of obtaining samples and cores. A decision will be guided by the extent to which information is to be quantified and the degree of finesse on such matters as the use of double- or triple-tube core-barrels for example. To what extent is the precise location of the borehole important and does this justify continuous records of inclination and azimuth? The orientation of rock cores may provide vital information in the interpretation of the geological structure.

Boreholes should extend to well below, say one diameter below, the invert of the tunnel, or deeper where the base of the geological facies chosen for the tunnel may be close to the tunnel and variable in level. If the maximum depth of the tunnel is not known at the time of the investigation, allowance must be made to guard against subsequent increase in depth placing the tunnel in unknown territory. The particular circumstances of the project will determine whether detailed sampling and testing is required throughout the length of the borehole or predominantly in the vicinity of the tunnel. The boreholes provide direct evidence from a minute fraction of the ground, probably transverse to the line of the tunnel; statistical validity of data may be improved by testing along a greater length of the hole through the same suite of rock. It is expensive to return to obtain more detailed information. There are many instances of tunnels being relocated during the planning process beyond the reach of the initial investigation.

The logging of rock cores should be undertaken as soon as practicable. The driller should have recorded all data necessary for the determination of the position of each core and of such matters as known loss of core, features of water entry and of any *in situ* testing. The engineering geologist will allocate the core stratigraphically and record features of engineering interest to an appropriate level of precision, prior to any part of the core being selected for testing (Geological Society of London 1970). Samples and cores from boreholes may be required to be maintained in good conditions over long periods, possibly protected from drying out or freezing. Soil or rock properties may be inter-related so that a readily observable feature serves as a marker for a more subtle feature of engineering significance, allowing zoning to be based on such identification.

The s.i. needs also to consider the consequences of tunnelling, particularly related to ground and water movements. The calculation of settlement (Section 5.3) may need special care for collapsing soils (e.g. loess) or for sensitive clays (e.g. marine clays affected by base exchange) or for made ground with high voidage (e.g. tunnelling beneath end-tipped embankments).

For major projects, or where experimental tunnelling practices are to be adopted, large diameter boreholes or test headings may be justified, allowing direct examination of the ground, the taking of *in situ* cores, possibly ascertaining the effectiveness of excavation or support systems. The transference of experience from such test headings to the underground project must make allowance for factors of size, differences in rock stress and water pressure, possibly for local differences in geological history.

Where the tunnelling allows periodical access to the face and to exposures of the ground along the tunnel, geological logging provides the most effective means to record the continuity and variability of relevant geological features. A common problem is that the mass of information recorded becomes unwieldy to synthesise (and may merely be used to compute one or other rock mass classification index). For the Channel Tunnel, the problem was eased by the use of a data base permitting selective examination of trends of particular features (Warren *et al.* 1996).

Geophysical methods of prospection have developed predominantly for the oil and gas industry, where their primary purpose lies in extending information obtained from boreholes into the second and third dimensions. The most widely used geophysical methods of prospection are those of seismic reflection and refraction. Each depends upon detecting the effects of differences of sonic velocity of continuous layers in the ground. Treating the ground as an elastic medium, the velocities of compressive and shear waves (C_p and C_s) are given respectively (Jaeger and Cook 1979) by:

$$C_p = [E(1 - v)/\rho(1 + v)(1 - 2v)]^{1/2} \tag{4.1}$$

$$C_s = [E/2\rho(1 + v)]^{1/2} \tag{4.2}$$

Seismic reflection techniques are most readily undertaken on water to ensure efficient transmission of the signal energy and are able to detect bands of sufficient contrast in sonic impedance of a width equivalent to at least half a wavelength. The higher the frequency, the greater the detail while the lower the frequency, for the same pulse energy the greater the penetration. Since the signal is not sharply focused, it is not possible to identify steeply inclined reflectors, which includes deep local weathering. Normally, a transmitter is attached to the survey craft which trails the receiver. Unless separate means are provided for recording the precise relative positions of each, the runs should be undertaken along, rather than across, any tidal current in order to preserve alignment between transmitter and receiver. Tie lines are also surveyed, preferably near slack water, so that the grid of records facilitates the tracing of individual reflectors. As with all such techniques there are methods for cleaning up the record, including the removal of multiple reflections, between strong surface and internal reflectors and the sea surface.

Seismic refraction depends on measuring the time of the return of the signal to surface receivers, and is only operable where there is increase of sonic velocity with depth; in a similar manner, electrical conductivity prospection requires increasing conductivity with depth. The calibration of such methods requires knowledge of the approximate numerical value of the appropriate physical characteristics on which each is based.

A technique with doubtlessly unexploited potential is that of sonic tomography, usually undertaken between boreholes, whereby the body of the ground may be explored in two or three dimensions by the interpretation of signals received at different depths in one borehole from transmissions from different depths in another. Large computing power is needed for the 'matrix inversion' required to provide results. The method lends itself to exploiting features of sonic velocity and of loss of signal strength.

Ground radar has been used for shallow prospecting and to investigate anomalies locally to a tunnel. Geophysical systems with directional capability with application to tunnelling, which make use of techniques comparable to those of remote sensing SAR (synthetic aperture radar) scanning from satellite, may well become available for ground prospection in the future.

The most detailed investigation of potential hazards immediately ahead of the tunnel is undertaken by probing. This may be of a selective nature, e.g. for the Heathrow Cargo Tunnel (Muir Wood and Gibb 1971), see Section 2.3. More generally, the need is for exploration around and ahead of the tunnel by means of an aureole of probe-holes inclined at an acute angle to the line of the tunnel. The features of such probe-holes in relation to different methods of tunnelling are described in Chapter 6.

4.1.3 Instrumentation and its interpretation

This brief account considers principles and not the characteristics of particular types of instrument. Instrumentation techniques have developed rapidly in recent years and may be expected to continue to develop in reliability, precision and reducing cost. Direct measurements of value to the engineer include:

1. measurements of movement of the ground and of structures, which may be used to derive strains; also, where appropriate, associated stresses may be calculated from strains;

2. measurements of ground-water pressures and levels, used for estab-
    lishing effective stresses in the ground, for determining hydraulic
    gradients and hence directions of flow, and, particularly, for measuring
    changes as a result of tunnelling.

Movements along the line of a borehole may be determined by measurement of changes of distance between indicator rings, by means of a probe
or by extensometer rods. Lateral movements in a borehole and angular
tilts of structures may be measured by inclinometer. Similar devices
may be used in probe-holes drilled from a tunnel or, alternatively, direct
measurement of relative movements of points anchored in the ground may
be measured at the tunnel face by multiple head extensometers.

Fluid-filled levels have been used to detect relative vertical movements
between points; it is then important to ensure no air in the system, no
differential effects caused by temperature or variation of density, no
prevailing atmospheric pressure gradient. The English Channel, as a large-
scale water-level, was used by Cartwright and Creese (1963) to establish
the relationships between the French and the English levelling datums
(IGN and ODN) using the electrical potential difference at the two ends
of a disused telegraph cable to record mass flow of water through the
earth's magnetic field, with correction for Coriolis effect, for atmospheric
pressure and wind gradients, leading to a difference in mean sea level
of around 80 mm, thus a difference between IGN and ODN of about
440 mm [subsequently corrected by GPS to about 300 mm, Varley *et al.*
(1992)].

Precise (invar) tapes or wires may be used to measure between pins
attached to the rock or to the lining of a tunnel. Alternatively the position of optical targets may be determined precisely by electronic distance
measurement from a theodolite. Stress levels in a tunnel lining may be
measured directly by jacks or a form of flat jack which transmits the load.
Qualitative measurement may be made by stress cells in a concrete lining,
or strain gauges on a metallic lining. Any direct measurement must consider
effects of stiffness in compression and in shear, relative to the part of the
structure it displaces, and whether measurements will be representative.
Drying of the face of a concrete tunnel lining sets up differential shrinkage
stresses which interfere with direct measurement of stress caused by ground
loading. Stress levels between the ground and the tunnel structure are yet
more difficult to measure. The device needs not only to have correct stiffness in compression and in shear but also to represent the surface of the
structure in position and in roughness. Records of the instrumentation for
the Severn Cable Tunnel (Haswell 1973), which appear to indicate high
out-of-balance forces, illustrate the problem. For a continuous ring, where
bending moments are derived from measurements of stress, missing data
may be estimated from the knowledge that:

$$\oint (M/EI)ds = 0 \tag{4.3}$$

The engineer is interested in deriving information from the synthesis of a series of different sources of data. The required precision needs to be considered at the outset. Synthesis of data of widely different accuracies may be misleading. For the purposes of research, interpretation may be a relatively leisurely process. The more immediate requirement will relate to control of the work of construction. For this purpose, particularly where measures are involved concerning not only the safety of the tunnel but also of structures and services, there will be large quantities of data which need to be continuously collated and collectively interpreted in real time in such a form as to be immediately useful to decisions affecting several simultaneous operations, with special concern for dealing with anomalous data which do not fit prediction. This is particularly the situation which requires complex data handling and correlation for the control of compensation grouting as a tunnel advances, described in Section 5.3.

Much surveying is undertaken by Global Positioning System (GPS) using satellites with local controls to obtain maximum precision. Traditional surveying was affected by gravity and hence a 'level' surface would follow the geoid and thus be 'correct' in relation to surface levels of still water. The GPS relates to geometrical coordinates of the earth and levels may therefore need to be adjusted in relation to the geoid for projects covering a wide area, particularly for water tunnels through mountains, affected by gravity anomalies, also possibly, over long distances, to compensate for the appreciable crustal movements caused by earth tides.

4.2 How not to manage the site investigation

As an aid in defining the qualities of sound s.i. for successful projects it is useful to reflect on some of the defects currently practised in this vital area. The ground is the principal determinant for tunnelling, its concept, the form of the permanent works, the manner and means for achieving this form. The site investigation and its sub-set, the ground investigation, represent a vital resource of great value to the project overall. In consequence it should follow that the greatest sharing of knowledge about the ground should occur among all the key participants, those who decide on principles and details of design and of construction. Indeed, experience of projects demonstrates that overall success correlates well with the degree of promulgation and sharing of data. Furthermore, in a related manner, much benefit follows from the application of relevant overall tunnelling experience to the design of the s.i. Otherwise, too often this becomes an exercise for demonstrating the esoteric knowledge of specialists in specific areas of the earth sciences, who do not understand the features, or combination of features, useful for interpretation in practical terms by the engineering geologist or geotechnical engineer for application to the needs of the tunnellers.

The ground investigation involves the investment of expertise, effort, time and expense commensurate to the size and value of the project, the nature and complexity of the geology. The results of the site investigation require specialist skills in interpretation which will relate the findings to knowledge acquired from other sources and to their interpretation. While the specialist attributes of those supervising s.i. are often emphasised, too rarely is emphasis also given to their ability adequately to understand the nature of the applications of the work to the tunnelling process.

It might appear axiomatic that the greatest care should be taken in ensuring that the s.i. addresses issues most vital to the project and that the fruits of s.i. and its interpretation, with whatever reservations about uncertainty may be advisable, should be made readily available to all those concerned in applying the results to the success of the project. How extraordinary, therefore, to find as common practices:

1. The engineer commissioned to organise and supervise site investigations is appointed by competition, does not need to display an understanding of tunnelling and is not otherwise engaged in planning or design of the project. This practice inhibits continuity or integration of project development, with the prospect that the s.i. will not be coupled with other features of the project planning process and will, in consequence, not provide adequate answers to vital questions and will not allow consideration of innovative methods of working.
2. Contractors engaged to construct the project are denied access to the interpretation of the s.i., allowed to view only the raw results of the s.i. which has been undertaken specifically for the project (c.f. good practice as described in Section 2.3). These data are made available for inspection only during a limited tender period and without guidance of discrimination as to what may be relevant among what are often large quantities of irrelevant data mixed with the vital information.
3. Tendering contractors are informed through the Contract Documents that the s.i. is provided by the Owner without warranty as to its accuracy, yet this may be the only source of information and it is usually quite impractical to supplement the data during a limited tender period. To deepen the inequity (and iniquity), the tenderers may be required to take full responsibility for all ground conditions without any entitlement to claim against unforeseeability.

How have these remarkable confusions of purpose and perversity of attitude come about and with what objective? What are the intended consequences? What are the actual consequences? We need to look no further than the law.

A commercial lawyer, consciously or subconsciously, imagines his client as engaged in a series of legal jousts, potential or actual, with all with whom he deals. The lawyer starts from the, apparently commendable, notion of protecting his client. A simple, alas far too simple, means is to pass responsibility for any conceivable source of risk down the line to all those (other than the lawyer) engaged to participate in the project. This however is totally contrary to the achievement of good engineering based on systematic control of risk, upon which depends the attainment of the Owner's objectives. The consequences may be briefly described in these terms:

1. The s.i., divorced from those who need to apply its results to the design of the process of construction, may neglect issues vital to the choice of the optimal scheme of construction.

2. Whatever may be the legal basis for the disclaimers as to the validity of s.i. data, the information provided thereby will often be the only, and certainly the most relevant, basis for the tender. The tenderers' interpretation may in consequence be partial and hurried. If the data are in any way misleading, unrepresentative or erroneous, particularly if such defects lead to apparent consequential impracticability of performance, there must, at least in equity, be potential grounds for claiming relief.

3. The cost of the project will be excessive for these direct reasons:
   • The Contractor, unless ready to accept loss, may assume the worst conceivable combination of geological circumstances, accentuated by lack of opportunity to undertake a deliberate considered assessment of all the data and thus unable to determine the optimal scheme of construction.
   • Absence of interactions between s.i. and other features of project planning and definition (see also Chapter 3) prevent optimisation.
   • Absence of adequate s.i. related to a specific means of construction may prevent this means, possibly the most appropriate, from being seriously considered by the Contractor.
   • Where the ground departs from expectation, the Contractor can expect no contribution from changes in the specified requirements for the Works which may well constitute a vital part of a practical technical solution.

4. Perhaps the most damaging feature overall is that the practice of using s.i. as an apparent weapon to emphasise the weakness of the Contractor in potential litigation, rather than a buttress to overall planning of the project, establishes a point of departure for distrust liable to poison relationships, leading towards adversarial positions, from which much else damaging to success may spring.

5. In the event of the Contractor encountering unexpected problems, unless readily mastered, there being no resource for compensation,

there must always be the temptation by one side to exaggerate the problems, possibly leading to their actual exacerbation, and by the other to pretend, having had to make no previous declaration of commitment to specific interpretation, that these were precisely what were to be expected (in France *pronostic rétrospectif*). These are not attitudes likely to lead to rapid economic recovery of the situation; in fact, they undermine the professional relationships between those concerned, on which success hinges.

The practices described above are particularly liable to lead to litigation, which only rarely provides satisfaction to those outside the legal profession. Additionally, if data available to the Owner or the Engineer or an interpretative report not issued to tenderers point even indirectly to any source of hidden danger, such as, for example, the possible presence of methane below the surface, without alerting the Contractor to this possibility, there might well be liability for concealing this potential risk if it were to eventuate. Is the above account a fictional scenario? Unfortunately not: each element is based on experience.

4.3 How much site investigation?

Since s.i. is accepted as being so vital to tunnelling, several attempts have been made to express the optimal expenditure on s.i. as some numerical relationship to the associated project, e.g. as a percentage of cost, or as a ratio of aggregate length of borehole to length of tunnel. Thus Legget and Hatheway (1988) suggest a range of 0.3–2.0% of total cost while West *et al.* (1981) suggest Stage II s.i. (Table 4.1) as representing 0.5–3.0% for the United Kingdom. For similar projects in similar ground there may well be merit in making comparison (bench-marking), but it is impossible to adduce rules of universal application for such reasons as:

1. Geological conditions between projects, and even within a single project, may be highly diverse.
2. Where the nature of the ground and the geological structure may be familiar from previous investigation – and particularly from previous tunnelling – the percentage cost of s.i. may be expected to be less than that for a similar project in previously unexplored ground.
3. A simple homogeneous depositional geology may generally be expected to yield requisite information from less s.i. than complex tectonically disturbed strata.
4. Where tunnels are set at considerable depth at expectation of saving cost, the associated cost of each borehole for investigation will increase exponentially (to a power well in excess of unity) with depth.

5. A s.i. campaign may lead to variation of the route of the tunnel with the prospect of increased cost of adequate investigation overall. Likewise, additional costs will be incurred when different tunnel options, in form or layout, are to be investigated.

6. A particularly economic form of tunnelling may be relatively intolerant of variations in the properties of the ground and may thus demand more and better quality investigation than a more tolerant or adaptable method of tunnelling.

7. Where special expedients (see Chapter 5) may be required, investigations of a specific nature may be needed to establish the efficacy of such practices and hence the selection of the optimal scheme.

8. With advances in technology both in s.i. and in tunnelling, one may expect to find historical changes in the ratios of costs.

9. Where trial shafts or tunnels are justified to help to establish optimal schemes of tunnelling, these may or may not be classified as s.i.

10. S.i. for an underwater tunnel will cost more than for a comparable tunnel under the land, partially compensated by the greater scope for underwater geophysical prospection.

It is undoubtedly true that certain tunnels have encountered problems that would have been eliminated by more effectively planned s.i. On the other hand, some of the most economic tunnelling in Norway (yet more economic by the value of aggregate obtained during excavation) through massive granitic rocks have required virtually no advance geological exploration on account of the general familiarity with the properties of the ground coupled with the tolerance of the drill-and-blast method of working, capable of variation in relation to conditions observed at the tunnel face.

The spacing between boreholes must take account of two different but inter-related criteria:

1. The extent to which the boreholes – possibly in association with geophysical prospection – are relied upon to define the presence or locality of particular features or of the variability in geological surfaces or horizons vital to the scheme of tunnelling.

2. The dependence on the boreholes to acquire general information about the qualities and variability of the rock which may contribute to a synthesis or a statistical analysis of value to the project.

In essence, the frequency of the boreholes must depend initially on Stage I of the studies (Table 4.1) amended as information begins to be assembled from the initial phase of Stage II.

A more general problem in assessing the adequacy of s.i. arises from the attempt to measure success of a project in terms of the percentage increment in tunnelling cost above the value of the Tender (corrected for inflation). This is highly unreliable for several reasons. Firstly, the Tender total may or may not include appropriate figures for contingencies; if it does not it does not represent the Tender value. Secondly, 'bench-marking' (see Chapter 6) between tunnel projects is notoriously difficult so there is no easy way of determining what a particular project *ought* to have cost. Thirdly, attempts to secure cost certainty for an uncertain project at the time of Tender will frequently be associated with elevated costs (see Section 3.7). Cost certainty at an early phase of the tunnel may well have appeal to the accountant on account of the high standard of cost control that this may appear to represent, but the benefit overall is illusory and the effect on cost control overall damaging, apart from the proclivity of such an approach to litigation. Curiously, the costs of litigation sometimes seem to escape from the total figure stated for the cost of a project; clearly such costs should figure in any comparison of costs overall.

Where the cost of a project has increased as a result of encountering an unexpected feature of the ground, it is too easy to conclude that the feature would have been revealed by a more thorough investigation. It is only by exploring how understanding about the ground has developed during the s.i. that it is possible to deduce whether or not supplementary s.i. might have been targeted to reveal such a feature.

The Kelvin sewer tunnel (Sloan 1997) appears to provide a simple example where excessive spacing between boreholes without intermediate geophysical data failed to reveal that the ice-eroded surface of the rock would fall below the crown of the tunnel. A more complex example is provided by the known presence of anomalies in the London clay, apparently associated with local erosion of the surface of the clay, fine-grained material from beneath being forced upwards by excess water pressure, leaving roughly cylindrical zones of unstable ground (Berry 1979). These features represent one of several aspects of periglacial phenomena described by Hutchinson (1991), encountered by several tunnels in the London area. In 1983, a 2.5 m diameter wedgeblock-lined water tunnel for the Three Valleys Water Committee, constructed between Wraysbury and Iver, to the west of London, encountered such a feature affecting a length of about 50 m of tunnel at a depth of more than 30 m. After unsuccessful attempts of control of the unstable ground by grouting, adopting the technique of 'claquage', i.e. fracturing the ground to assist penetration, with sodium silicate grout, freezing from the surface by the use of liquid nitrogen was used successfully.

A major crisis in the construction of the North Bank machine hall for the Kariba hydro-electric project depended in part upon the absence of identification in the site investigation of bands of biotite schist in the gneiss rocks in which excavation was undertaken (Anon 1974). Differences between experts upon the significance of these oversights persisted as the original contractor was expelled (and went into liquidation as a result) and the work completed late and at an increase in cost of around 50%.

It needs to be understood that, by failing to disclose a suspected unwelcome feature of the ground, a borehole does not establish its absence. This is particularly relevant for example to the deep tropical weathering of igneous rocks. The weathered zone may be fairly narrow, following a fault and reducing in width with depth. In consequence, the probability of it being encountered by a series of boreholes will depend on:

- the spacing between boreholes in relation to the spacing between the weathered zones;
- the orientation of the zone in relation to the inclination and azimuth of the borehole.

Furthermore, where the width of the zone may be expected to reduce with depth, the probability of an encounter by a borehole will also reduce with depth. Thus, the series of boreholes will indicate a *minimum* depth of weathering but cannot establish the *maximum* depth. Sonic tomography between boreholes, or the use of steered boreholes might provide such information for a particular feature, providing that the suspected position is reasonably predictable from surface feature or other evidence. A statistical analysis of the data may help to establish the relationship between depth of the tunnel and its likelihood of encountering the feature.

From the above it should be evident that the needs of a project are highly site- and project-specific. Once the fundamental information is obtained to permit particular forms of construction to be considered, questions of uncertainty and how these may be effectively reduced, must dominate the strategy for further s.i., the criterion being that the cost of additional s.i. must more than compensate the value of the expected reduction in cost of construction, which of course will include the cost of uncertainty. Apart from exploration and 3-D mapping of the ground, as the planning of the project unfolds the supplementary s.i. may also be required to investigate the practicability of specific techniques, including special expedients (Section 6.4). These may well entail specific trials designed for the particular application in representative ground.

4.4 Reporting on site investigation

Reporting on s.i. needs to respect the cadence of the programme for project planning and design (Table 4.1). Essentially, a summary is required of Stage I, listing with care the source of each element of data. A report will be necessary after each phase of Stage II and a specific intermediate report on every other occasion at which new interpretation of data might lead to a variation in the *design* strategy.

Once again, emphasis is placed on the need for those in charge of s.i. to combine an understanding of the essential features of geology with an appreciation of the significance of the combination of the features in relation to the problems and opportunities of tunnelling. This combination is particularly important in enhancing the ability to draw upon the significance of the state of knowledge of the geology at an intermediate stage of the investigation. It is just as important to appreciate the extent of uncertainty in a particular respect as it is to demonstrate ascertained facts and their possible or probable inferences. For example, where figures of permeability have been derived from packer tests in boreholes (Appendix 5G) consideration should be given to such features as:

1. explaining the technique involved;
2. providing full records of measurements in tabulated figures and graphical form;
3. interpreting figures of permeability, with indication of reliability related to the nature of the test, the shape of the pressure/flow curves (Muir Wood and Casté 1970) and other material factors;
4. another reminder that values of permeability are vertically skewed, i.e. that measurements are predominantly of horizontal permeability.

In relation to point 4 above, for measurements of permeability of unweathered chalk for the Channel Tunnel, where hydraulic conductivity depends predominantly on fissure flow, permeability factors between 2 and 5 are recommended to infer vertical permeability from figures derived from packer tests in boreholes (Sharp *et al.* 1996). As this book goes to press, results are awaited of a numerical study of this feature.

The logging of cores should follow recommended procedures, in Britain those of the Geological Society of London (1970). The commonest way of recording the degree of jointing of rocks is by way of Rock Quality Designation (RQD) as defined by Deere *et al.* (1967) as the percentage of the total length of core recovered in solid pieces greater than 100 mm in length. Stereoplots (Figures 4.2 and 4.3) may assist in identifying dominant trends of jointing and critical needs for tunnel support. Study of a suite of rocks may permit rock strength (q_u) to be related to elastic modulus (E) as illustrated by Figure 4.4 as a first guide to support needs.

There is a logic in the progression from geological description, working from the general towards the particular for the project, through the expression of features in quantified terms as engineering geology, thence to geotechnical engineering and its application to the planning and design of the project, of its elements, how they are to be built and by what means.

There is however considerable variety in application of such a 'linear' approach, for reasons explained in Chapter 2. Innovative approaches to

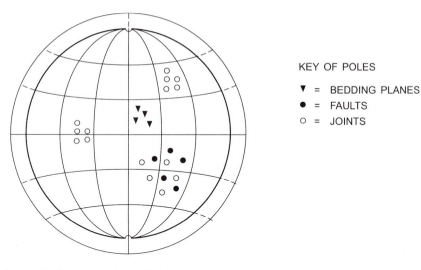

KEY OF POLES

▼ = BEDDING PLANES
● = FAULTS
○ = JOINTS

Figure 4.2 Stereo-plot for rock joints.

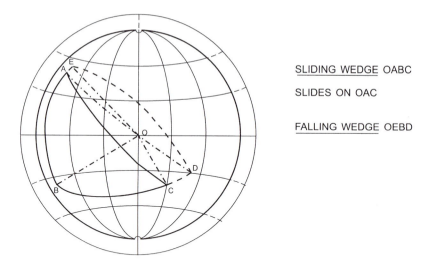

SLIDING WEDGE OABC

SLIDES ON OAC

FALLING WEDGE OEBD

Figure 4.3 Stereo-plot for unstable rock blocks.

any aspect, or incidental surprises in geological interpretation, may lead to iteration or re-examination of an earlier phase of the work. A more fundamental issue is that of variability of the ground and the identification of the varying features. Optimal tunnelling in relatively homogeneous soft ground, for instance, may be based on highly developed 'ground

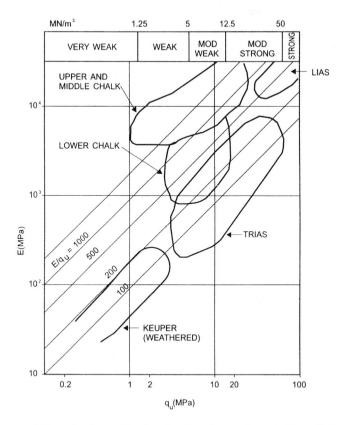

Figure 4.4 Typical relationships for *E* and *q~u~* for weak rock (after Hobbs 1974).

models' (Appendix 5F), which at the extreme may be expressed in the form of constitutive equations for the behaviour of the ground. Rock tunnelling in squeezing ground lends itself to a different form of analytical approach, with variability providing limits for sensitivity analysis, i.e. the testing of a design against its capacity for variation against possible ranges of parameters, controlled by an observational technique. For tunnelling in jointed competent rock, the main objective of the site investigation will be to determine the degree of variability of the jointing pattern and of the associated problems with inflow of water.

Informal Support will be adaptable to demand so there is less need to establish a highly specific ground model on which to base the requirement for support, providing that modification may be secured within the time available, and time-based behaviour may then become an important issue to be explored by the s.i.

In summary, rock tunnels require:

- an account of the variability in support requirements, sufficiently quantified, the data graded in respect of quality, to permit realistic estimates of cost of different solutions;
- data necessary for the design of methods of excavation, whether by machine or by drill-and-blast;
- prediction of specific problems to be expected, in nature and, so far as possible, extent and expected locality.

The latter is frequently the most important, but overlooked, factor. The economics of a scheme of tunnelling advancing at 100–1000 m/month may be utterly compromised by a 6-month delay to deal with a single unforeseen, but foreseeable, localised unstable water-bearing feature. There are many problems along the way in identifying such a feature:

- the intersection of a disturbed zone by a borehole may be recorded by the laconic 'loss of core';
- undue reliance on surface expression of features, particularly in a dry climate where induration of faulted material at the surface may give a totally misleading impression of its nature at depth;
- surface mapping may be the most positive means for locating and orientating features of concern but local weathering may obscure such exposures so that more subtle secondary features may need to be identified.

The Canyon Project, a 100 MW, 540 m head, hydro-power project of the Upper Kelani Valley basin of Sri Lanka was commissioned in 1982 at a time of considerable power shortage. The original alignment of the penstock tunnel was found to be following a band of karstic limestone metamorphosed to marble and altered to clay with boulders, pebbles and sand, which might have been foreseen from a surface line of swallow-holes and boulder-filled depressions (Vitanage 1982). The tunnel was in consequence diverted but ran close to this feature for about 80 metres and traversed it obliquely over a further 50 m. Considerable quantities of cement were used for cavity grouting.

On first filling the tunnel, damage was apparent, evidenced by leakage at the rate of about 30 litres/s. On dewatering, sand and silt were found as a deposit in the tunnel. There was evidence that a length of the crown of the tunnel had been displaced (opening 3–10 mm, shearing 3–10 mm, but possibly greater under pressure). The tunnel was repaired and regrouted. In view of a prevailing power shortage, the project was recommissioned. The question was then posed as to the temporary policy prior to undertaking major permanent remedial work. The advice given was to avoid surges, to maintain vigilance for any appearance of silt at the power-house and to sink two boreholes into the altered geological feature in the vicinity of the tunnel. By continuous observation of water levels, sudden rise or fall would signal a change in the local regime requiring further investigation. Meanwhile, the plant could continue to operate. At intervals when power was not required, the rate of leakage out of the tunnel could also be observed.

Directional skewing occurs in the measurement of rock properties, such as RQD, measured in boreholes. Results should always be qualified in such respects, particularly where there are suspicions of anisotropy. Anisotropy is a general rule rather than an exception for several reasons:

1. Variations in patterns of deposition, erosion and the complex effects of periodical surface exposure will tend to cause layering of clastic strata (e.g. seasonal varving) with consequential effects on properties as explained in Appendix 5F.
2. Many of the causes of jointing and folding in rock will tend to cause sets of vertical or sub-vertical joints and of jointing parallel to the bedding. Weathering, associated with ground-water flow, may tend to follow the jointing pattern, hence the phenomenon of deep local weathering in igneous rocks and karsticity in dolomitic limestones. Intrusive rocks will tend to follow pre-existing joints orthogonal to the direction of low ground stress. Vertical boreholes are obviously ineffectual in providing statistically reliable information on the frequency, vertical extent and significance of vertical or sub-vertical features.
3. Schistosity will be orientated in relation to the stress regime in the ground at the time of metamorphosis.

The most important feature of a borehole may concern the causes for loss of core recovery. Down-the-hole logs and cameras may help to explain their significance.

4.5 Identification of patterns in the ground

Site investigation will provide direct evidence of only a minute fraction of the ground to be penetrated by the tunnel. Indirect evidence by geophysical means may well provide valuable information of likely continuities and discontinuities, interpreting structure in relation to specific features, with the prospect of correlations between such features and properties of interest to tunnelling.

Too often, results of a site investigation are then presented in the form of a graphical plot and a statistical analysis for a particular rock type or suite of rocks. Much may be gained by searching for patterns in the data which may assist in predicting the nature and variability of the ground to be encountered by the tunnel. Patterns are of several varieties, of which the most interesting for tunnels are:

1. patterns of deposition (and possibly of local erosion), including cyclical variation and trends along an individual clast;
2. patterns of change by pressure, temperature, alteration (preferring the French term to the more confined implication of 'weathering'), water flow, diagenesis;
3. patterns of effects of tectonic activity in relation to the physical characteristics of the rock.

From the several phases of investigation for the Channel Tunnel a number of these patterns became evident. Cyclical variations in the chalk marl of the Lower Chalk were associated (Harris *et al.* 1996) with alternating layers of more and less clay/carbonate ratios. The overall variation in properties along the tunnel could be largely explained by an increasing thickness of the clay-rich layers towards the English coast, with an accompanying overall increase in thickness of the Chalk Marl of the Lower Chalk (Mortimore and Pomerol 1996). There were also patterns of decreasing permeability with depth, largely associated with increasing clay content. A further pattern was related to the extent to which varying sea-level through geological time subsequent to deposition had encouraged flow paths to develop along fissures and joints. A related pattern concerned the consequences of tectonic activity in its effects of anticlinal and synclinal folds, with wrench faults, on the rock properties and on the development of faulting. We may expect to be able in the future to develop improved means for relating the effect of tectonic activity at depth to the consequences to overlying rocks of interest to tunnelling (see, for example, Varley 1996), which could be of considerable value in predicting variation in rock quality along the line of a tunnel.

Episodes of interruptions in deposition may be associated with surface weathering but more often with erosion and variation in deposits which may be traced as part of the pattern, appreciating that one erosional episode may lead to the partial or complete removal of the evidence of earlier episodes of the same, or of a different, nature. This may result for example in the presence of confined aquifers, as lenses of sand in ground of otherwise low permeability, which may avoid detection by normal site investigation.

Intrusive igneous rocks, basalts and dolerites, may themselves be susceptible to deep weathering. They may also give rise, on cooling, to the development of contraction joints, providing flow routes for water and hence increasing the local susceptibility for further weathering.

Study of the ground-water flow may establish patterns of chemical/ physical erosion and of chemical deposition, leading to the sealing of water channels.

Alteration at depth away from intrusive igneous rocks is most likely to be associated with ground-water movements and hence to the pattern of jointing and to the historical episodes causing forcing of ground-water flow. The rate of change will be related to the temperature and chemical content of the circulating water. It is important to understand the patterns of change of relative sea level which may determine the levels to which major weathering has occurred.

If patterns are expected to be identifiable in the ground to be investigated, techniques need to be adopted for their identification and mapping. For example, palaeontology may assist in relating a feature of a pattern to a particular geological horizon and hence to tracing it through the ground (Bruckshaw *et al.* 1961). The most positive means of identifying and delineating patterns may be by the use of tomography. Such techniques are usually excessively expensive for normal tunnelling but may well become the norm for specialised application such as exploration for underground nuclear waste disposal or for caverns for gas or oil storage.

When undertaking initial geomorphological prospection, the identification of bands of rock resistant to weathering will often help to explain some of the less obvious topographical features. Similarly, the spring line at periods of raised water-table may be identified from the shape of valley features and hence to the approximate location of the levels of aquicludes. Many other similar associations may be developed by the observant geomorphologist.

Ultimately, from the viewpoint of the tunneller, the objective must be to identify patterns of grades of engineering properties of the ground and in this way to predict rock 'zones' to be used for designing patterns of support, for predicting rates of progress and possibly as a basis for payment. Such zones depend on complex associations of rock mass properties, internal stress patterns, ground-water patterns, tectonic activity. During site investigation, thought needs to be given to the extent to which useful information leading to such predictive zoning may be acquired from design and synthesis of the recovery of data.

4.6 Specific features of site investigation

The first essential for successful site investigation for tunnelling is that of good communication between those concerned with the performance of the work and those concerned with its application across all aspects of the tunnelling process. The communicators may include engineering geologists – and their specialist advisors in particular aspects of geology and geophysics, geotechnical engineers, and designers of the tunnelling works with familiarity across the construction process, possibly advised by the designers of specialist processes and plant. At the outset it will generally not be possible to foresee which specific aspects of the properties of the ground will be most relevant to tunnelling. Periodical briefing between those illustrated by Figure 2.3 should ensure that these aspects are systematically clarified and the site investigation directed increasingly towards the elucidation of areas of important uncertainty. There needs to be similar two-way communication between the engineering geologist and specialists to whom he looks for advice and for whose work he is responsible.

The director of the investigation should always understand the current expectations concerning the ground. As Glossop (1968) has wisely stated: 'If you do not know what you should be looking for in a site investigation, you are not likely to find much of value.' With coherent and updated briefing, the discovery of an unexpected feature may lead immediately to question the prevailing 'ground model'. The term 'ground model' is used in two different senses:

- a specific model used for analysis of tunnel stability (Appendix 5F);
- (as here) a general descriptive statement of the ground, including features relevant to tunnelling, arising from synthesis of data currently available.

The planning of the site investigation depends on the second definition, the first being derived from it to provide a simplified basis for tunnel design.

The investigation may also need to explore the interaction between the proposed tunnel and other existing or proposed surface or buried features. In the most demanding circumstances, the s.i. may need to be augmented by detailed analysis or centrifuge model in order to determine safe tunnelling procedures. Data on characteristics of the ground will then be required in a form to feed into the constitutive equations (i.e. those equations which relate the physical characteristics of the ground affecting time-dependent relationships between stress and strain) needed for valid numerical analysis. Such work needs to be conducted with great discrimination. Much worthless numerical modelling of tunnels has been undertaken where the circumstances do not warrant such an approach, where the model does not adequately represent the construction process, or where the lack of specific data renders the results unreliably precise or positively misleading.

Traditionally, the phases of the site investigation (Table 4.1) have been discussed in relation to the phases of the contract for construction, i.e. phases prior to invitation to tender, possible interaction phase with tenderers (permitting for example activities arising from direct access to test shafts and headings), phases after award of Contract.

In this book, one objective is to insist that the development of the *design* process is the significant feature to which phases in s.i. must relate. For example, an early commitment to a particular means of tunnelling may permit an approach to a project whereby s.i. may be specifically tailored to this means from an early phase. The results of s.i. are not to be treated as a weapon of defence of a law-infested client but as a vital resource to be applied to the benefit of the project, with areas of uncertainty requiring at all times to be identified with as much emphasis as the areas of confident knowledge.

1. Gjøvik sports cavern for 1994 Winter Olympics in course of construction. The main cavern has a span of 61 metres, height of 25 metres (photo © Scan-Foto Hans Brox).

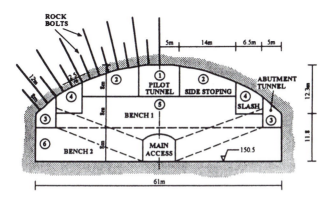

2. Cross-section of Gjøvik cavern indicating the scheme for construction.

3. Station construction of the Jubilee Line extension (photo courtesy of QA Photos Ltd).

4. Robbins 5 metre diameter TBM with back-up ready to start 17.5km drive for the 45km long Lesotho Highlands transfer tunnel (photo courtesy of QA Photos Ltd).

5. Installation of waterproof membrane behind in-situ lining of 250 metre long x 18 metre high Baixo-Chiado station cavern (photo by José Carlos Aleixo).

6. Formwork for the in-situ concrete lining for the Lesotho Highlands transfer tunnel (part of the 82 km total length of the tunnel) (photo by Bogdan Onoszko).

7. An Atlas Copco Rocket Boomer 353 ES drilling jumbo used for the Ullbro tunnel (photo by Shani Wallis).

8. Typical completed tunnel for the Jubilee Line extension (photo courtesy of QA Photos/Jubilee Line Extension Project).

9. Primary shotcrete support by top heading with two benches for approximately 14 metres diameter, North Downs Tunnel, Blue Bell Hill, Kent, for Channel Tunnel Rail Link (photo by Ros Orpin, and courtesy of Rail Link Engineering).

Chapter 5

Design of the tunnel project

Design – the continuous thread.

5.1 Options in tunnel design

5.1.1 The nature of the ground

Chapter 1 describes some of the developments in tunnelling which have contributed to the options available at the present day. For any specific project, the selection must be made against the known and suspected features of the ground, also of other relevant aspects such as access and local availability of tunnelling traditions and skills. The method needs to be considered in relation to tolerance or adaptability in respect of the variability of the ground (see Chapter 6).

Traditionally, in the days of simpler tunnelling techniques described in Chapter 1, the ground was subdivided, for the sake of defining the approach to be made, into 'rock' and 'soft ground'. Now that it is possible to learn, where appropriate, considerable information about the ground, not only in descriptive terms but also in behavioural terms, e.g. features associating stress and strain in a time dependent manner, possibly the basis of the ruling constitutive equations, the subdivision from the viewpoint of tunnel stability is more fundamental:

1. Ground to be treated as a continuum, i.e. all forms of soil and incompetent rock. Incompetent rock is defined as rock which will naturally and rapidly deform to close an unsupported cavity (say, $R_c < 2$ where R_c is defined below). The question of time is important since all rocks deform with time (although rates of deformation of familiar rocks at normal pressures and temperatures are imperceptibly slow), salt bodies being amongst the most readily deformable.
2. Ground to be treated as a discontinuum, i.e. rock whose behaviour is dominated, in relation to tunnel stability, by movement along joints between discrete blocks.

A: DESIGN BASED ON EXPERIENCE
B: RATIONAL DESIGN WITH OBSERVATION

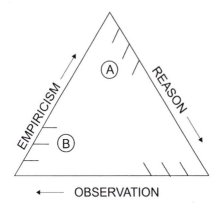

OBSERVATION

Figure 5.1 Factors contributing to conceptual design.

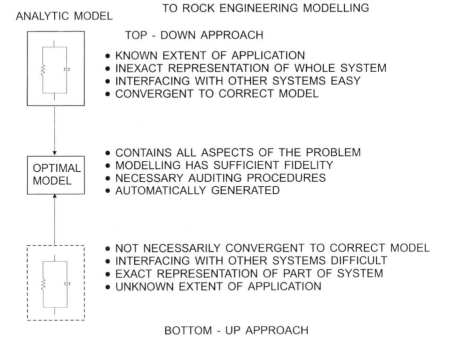

THE ANALYTIC AND SYNTHETIC APPROACHES
TO ROCK ENGINEERING MODELLING

ANALYTIC MODEL

TOP - DOWN APPROACH
- KNOWN EXTENT OF APPLICATION
- INEXACT REPRESENTATION OF WHOLE SYSTEM
- INTERFACING WITH OTHER SYSTEMS EASY
- CONVERGENT TO CORRECT MODEL

OPTIMAL MODEL
- CONTAINS ALL ASPECTS OF THE PROBLEM
- MODELLING HAS SUFFICIENT FIDELITY
- NECESSARY AUDITING PROCEDURES
- AUTOMATICALLY GENERATED

- NOT NECESSARILY CONVERGENT TO CORRECT MODEL
- INTERFACING WITH OTHER SYSTEMS DIFFICULT
- EXACT REPRESENTATION OF PART OF SYSTEM
- UNKNOWN EXTENT OF APPLICATION

BOTTOM - UP APPROACH

SYNTHETIC MODEL

Figure 5.2 The synthetic and analytical approaches to rock mass modelling (after Hudson 1993).

Occasionally, competent massive rock is so little jointed as to present no stability problem. More taxing of ingenuity is the jointed, relatively weak, rock whose treatment needs to consider the rock simultaneously as a continuum and as a discontinuum. Categorisation is affected by the relationship between tunnel size and joint spacing (Hoek and Brown 1980).

Where choice exists, the position, orientation, shape and direction of tunnel construction should take account of the rock structure and *in situ* stress tensor. Interbedded rocks, e.g. alternating mudrocks and silt-stones, may appear to form a continuum but the effect of changed stress patterns as a result of tunnelling may induce cleavage along bedding planes with local fracturing and instability. The special needs for support, or for adaptation of tunnel profile, should be considered for such circumstances.

Whatever may be the features of the tunnel to be designed, the approach should include these elements:

1. *Experience*, incorporating features of empiricism based on an understanding of ground characteristics and on successful practices in familiar or similar ground.
2. *Reason*, using analytical solutions, simple or more complex as the situation may demand, based on an acceptable 'ground model' (see Appendix 5F).
3. *Observation* of the behaviour of the tunnel during construction, developing into monitoring with systematic predesigned modification where a feature of Observational Design (Section 2.7) is to be adopted.

This approach is illustrated by Figure 5.1, recognising that points 1, 2 and 3 are complementary, contributing to the optimal approach to the particular circumstances rather than competitive techniques of tunnel design. However, the mix will depend greatly upon the circumstances. For a traditional form of tunnelling in familiar ground, reliance on experience will predominate (zone A in Figure 5.1); the design of a tunnel in a particularly sensitive area will depend upon a reasoned design combined with the adoption of the techniques of Observational Design (Section 2.7) and will thus lie nearer zone B of Figure 5.1. Hudson (1993) illustrates (Figure 5.2) an approach towards a 'ground model', with the objective of striking the greatest degree of proximity between the 'top down' and the 'bottom up' models, discussed in Appendix 5F.

Where a specific problem of stability lends itself to analysis, resort may be made to the Limit Theorems of the theory of plasticity (confusion may be caused by those who use the expressions 'upper bound' and 'lower bound' more loosely to define high and low estimates of the measures needed to ensure stability or safety):

1. *Upper Bound Theorem*: If an estimate of the plastic collapse load of a body is made by equating the internal rate of dissipation of energy to the rate at which external forces do work in any postulated mechanism of deformation of the body, the estimate will be either high or correct.
2. *Lower Bound Theorem*: If any stress distribution throughout the structure can be found which is everywhere in equilibrium internally and balances certain external loads and at the same time does not violate the yield condition, those loads will be carried safely by the structure.

Davis *et al.* (1980) have approached the stability of the face of a tunnel in clay, in circumstances in which the region of uncertainty between the upper bound and the lower bound could be acceptably constrained.

Many authors make a distinction between a 'passive' support system, in which the ground load gradually and naturally comes on the support, and an 'active' system, such as rock-bolting, in which the support is stressed on installation against the ground. There is no sharp distinction, however, since rock-bolts for example may be emplaced with or without initial pre-stress, depending on the system of anchorage and on the optimisation of strain in the rock, and may be combined with passive features such as sprayed concrete.

For tunnels in strong rock, i.e. where the competence ratio R_c (ratio of unconfined compressive strength of rock to initial state of stress in the ground, see Muir Wood (1972)) may be represented as, say, $R_c > 4$, the structural design process is predominantly concerned with the effects of discontinuities, generalised as jointing. As described above, the composite geometry of joints and the conditions of joint surfaces are therefore the major characteristics relating to the stability of the tunnel. Scale is of the utmost importance in relating the spacing of the joints or the sizes of potentially unstable blocks (Figure 4.3) to the cross-sectional dimensions of the tunnel, affecting the risk of potential modes of failure. Providing that rock joints are tight and that there is no risk of water dissolving joint infill, rock bolts will usually suffice to secure potentially unstable areas of strong rock or, by patterned bolting for more highly jointed rock, in combination with steel mesh or equivalent to create a self-supporting rock arch or ring (Appendix 5D). Where jointing may be described as forming significant 'sets', i.e. preferential jointing confined to a number of well-defined directions (Figure 4.3), the geometry of blocks which might become detached by falling or sliding into the tunnel may be predicted.

Where a joint of shear strength: $\tau = c + \sigma_n \tan \phi$ (appreciably less than the unconfined compressive strength of the unfractured parent rock q_u) occurs, potential slippage may occur for a limited range of angles of incidence, α, of the joint to the exposed rock face, as illustrated by Figure 5.3. The expression for the limiting circumferential unconfined compressive rock stress, σ_θ, is readily found, in terms of τ and of σ_r, the radial confining stress, by way of the Mohr diagram, Figure 5.4, to be the lesser of the unconfined strength of the intact rock and the value of σ_θ given by:

$$(\sigma_\theta - \sigma_r)\sin(2\alpha - \phi) = [2c\cot\phi + (\sigma_\theta + \sigma_r)]\sin\phi \tag{5.1}$$

At the periphery of the tunnel, $\sigma_r = p_i$, the support pressure. Where $\sigma_r = 0$, eqn (5.1) simplifies to:

$$\sigma_\theta = 2c\cos\phi/[\sin(2\alpha - \phi) - \sin\phi] \tag{5.2}$$

From the Mohr diagram (Figure 5.4) it is immediately apparent that the minimum value of σ_θ given by eqn (5.2) is:

$$\sigma_{\theta min} = 2c\cos\phi/(1 - \sin\phi) \tag{5.3}$$

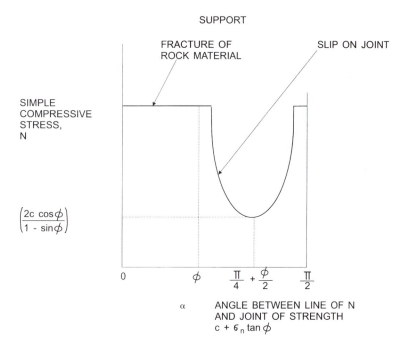

Figure 5.3 Failure criterion for jointed rock subjected to simple compressive loading.

for a value of α given by:

$$\alpha = \pi/4 + \phi/2 \tag{5.4}$$

Whether slippage along such a joint leads to a mechanism for failure and hence to a displaced block depends on the geometry of the jointing system in relation to the tunnel. Hoek (1983) describes experimental studies of the directional strength of a number of jointed rocks.

For large caverns, the particular risk is associated with major systems of jointing combining to define a massive potential unstable wedge or pyramid of rock, for which major stabilisation works with anchor cables, possibly combined with anchor blocks formed in special headings, may provide the solution.

For tunnels in weak rock, i.e. where $1 < R_c < 4$, or where the rock is so heavily jointed that stability of individual blocks is generally suspect, support will be required close to the face and this will dominate the design approach. For support needs, the provision depends *inter alia* on initial rock stress and hence on the depth of the tunnel, with special consideration given to localities where the ground surface is steeply inclined or where the major principal stress is not vertical. For soils, the at-rest ratio of horizontal to vertical stress $K_0 = \sigma_h/\sigma_v$ is usually considered as a constant for any specific point, but often varying with depth. For rocks, horizontal

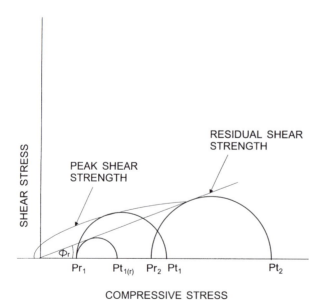

Figure 5.4 Mohr's diagram for rock.

stress may vary between minimum and maximum values orthogonal to each other. The major principal stress may be at any azimuth or inclination, depending on the geological history. At depth, continental trends in a dominant direction of high horizontal stress may be an important factor in design. Ward (1978) describes practical measurement and analysis of support needs for weak rock, largely based on studies for the Keilder Tunnel (Coats *et al.* 1982).

For soils and weak rocks stability at and near to the face will be time-dependent, leading, for an unsupported face, to the imprecise notion of 'stand-up time', which may be compounded by a number of factors:

- equilibration of pore-pressures for cohesive soils and weak rocks;
- changing stress pattern as the distance between the face and the last supported length of tunnel increases;
- discontinuities in the soil (e.g. slickensides, joints in clay, sand partings in silt);
- time-dependent effects of concentrated loading between blocks (of weak rock).

The effect of the equilibration of pore-pressures needs explanation. As a tunnel advances in clay, the reduced total stress in the ground close to the face, and the associated increase in shear stress, leads to a tendency for dilation of the soil which causes a local reduction of pore-pressure (possibly below zero on account of capillary forces). This reduction in turn increases the hydraulic gradient which draws water through the ground towards the tunnel. The reduced pressures cause a corresponding increase in effective stress of the soil, which is then largely responsible for the stand-up time. For an advancing tunnel, a 'steady state' occurs in the stress pattern relative to the face, advancing in a Lagrangian fashion with the progress of the tunnel (Figure 5.5). In consequence one expects stand-up time to increase as the rate of advance increases, but this depends essentially upon the coefficient of consolidation (or swelling) c_v, where

$$c_v = k/m_v\gamma_w \tag{5.5}$$

Here k = hydraulic permeability, m_v = coefficient of volumetric compressibility, and γ_w = unit weight of water. There is considerable evidence of the increased likelihood of instability of a stalled tunnel face.

Figure 5.6 indicates diagrammatically the features of supported pressure around a circular tunnel and of ground convergence in the vicinity of the advancing face of the tunnel. Ahead of the face of the tunnel, the 'core' (i.e. the ground within the cylinder to be excavated) will be more or less distorted by contraction across the diameter coupled with longitudinal extrusion towards the face. Dominant controlling features will be

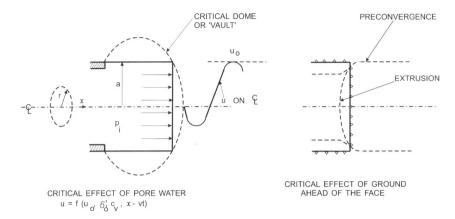

Figure 5.5 Features affecting stability at the face of a tunnel in clay.

the initial stress regime and the stress/strain characteristics of the ground. Panet and Guenot (1982) propose:

$$\sigma_a = (1 - \lambda)\sigma_0 \qquad (5.6)$$

where $1 > \lambda > 0$, to relate radial stress at radius a to initial stress, σ_0, in the ground.

For elastic ground,

$$\lambda_x = u_{a(x)}/u_{a(\infty)} \qquad (5.7)$$

where u represents convergence, x a distance along the tunnel, and ∞ relates to a distance remote from the face for a (theoretically) unsupported tunnel. In general, the greater the value of λ, the greater the convergence and the lower the required support pressure.

In the vicinity of the face, the stress patterns may be considered as:

1. a dome (i.e. part of a 3-D sphere or ellipsoid) of stresses bearing on the last effective length of ring of support or lining of the tunnel and on the 'core';
2. circumferential stresses around the unsupported length of the tunnel, mainly within the 'dome';
3. stresses parallel to the axis of the tunnel, whose divergence will add to the radial stressing of the ground ahead of the face and whose subsequent convergence will aid stability between face and tunnel support.

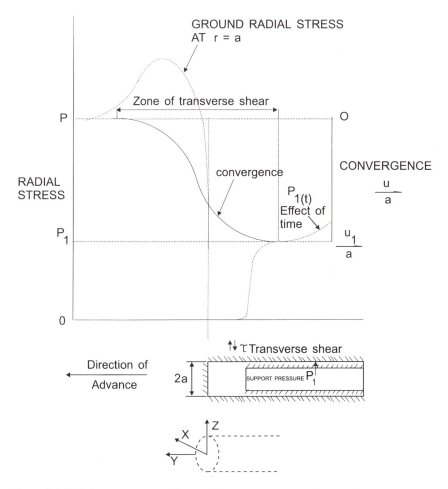

Figure 5.6 Radial convergence and ground stress in vicinity of tunnel face.

When stability of slices transverse to the tunnel are considered, the stress dome contributes to shear (in the plane r,θ) between slices.

This simplified concept of stress patterns indicates the contribution of the core and of early provision of a completed ring of support to achieve initial stability. If the 'dome' of ground support is viewed in a static manner, it might be supposed that the most recently completed ring of support could bear a disproportionately high loading. Since, however, the tunnel is considered to be advancing steadily, the incremental load on each ring cannot correspond to more than that due to the corresponding length of reduced support from the advance of the face, taking account of the contribution from increased circumferential stresses in the ground around the tunnel. Furthermore, the support is not provided instantaneously – there is a time for erection and a time for the support (e.g. shotcrete) to develop a high enough effective modulus – and the first effect of loaded support will be a degree of radial confinement of the ground such as to enable the ground to accept a greater share of circumferential load than it could prior to this confinement. As a consequence, the load on the support will increase as it recedes from the advancing face, affected by time and by the reduction in assistance from the third dimension locally to the face.

An appreciation of the 3-D arching or 'doming' phenomenon is essential to the strategy of supporting the ground. In weak (but not squeezing) rocks, the major component of support will occur around the tunnel with the longitudinal component of arching providing local support only. In soils, the longitudinal component of 'doming' assumes greater importance in magnitude and in the criticality of timing. Any 2-D figure of convergence-confinement, such as Figure 5.7, should therefore be understood as a gross over-simplification of the critical circumstances near the tunnel face where prompt decisions need to be made. Figure 5.8 (after Kidd 1976) illustrates application to the control of support needs for the Orange-Fish Tunnel.

For soils, it is customary to express the major factors affecting stability of the face in terms of the stability ratio (or 'simple overload factor') N_s based on Broms and Bennermark (1967) defined as:

$$N_s = (\gamma z_0 [+ q] - p_i)/c_u \qquad (5.8)$$

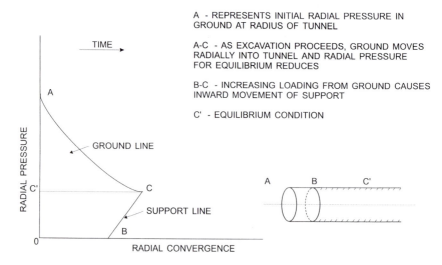

A - REPRESENTS INITIAL RADIAL PRESSURE IN GROUND AT RADIUS OF TUNNEL

A-C - AS EXCAVATION PROCEEDS, GROUND MOVES RADIALLY INTO TUNNEL AND RADIAL PRESSURE FOR EQUILIBRIUM REDUCES

B-C - INCREASING LOADING FROM GROUND CAUSES INWARD MOVEMENT OF SUPPORT

C' - EQUILIBRIUM CONDITION

Figure 5.7 Ground/support interaction – conceptual diagram.

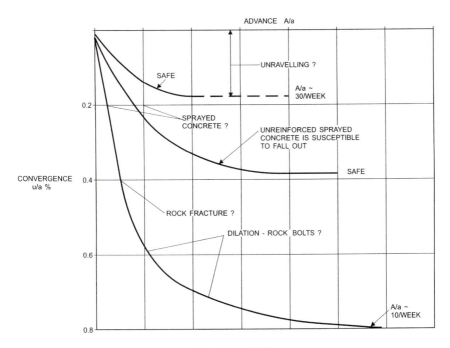

Figure 5.8 Convergence guide-lines (after Kidd 1976).

where (see Figure 5.9) γ = unit weight of soil, q = surface surcharge pressure (if any), p_i = internal support pressure in tunnel, c_u = undrained shear strength of soil, and z_0 = depth to axis of tunnel. It will be noted that, for zero surcharge and internal support, $N_s = 1/2R_c$.

Davis *et al.* (1980) have related stability of an advancing tunnel without face support to N_s for varying values of z_0, tunnel radius r and l, the distance from the face to effective support, supported by centrifuge model tests reported by Mair (1981). For values of $N_s < 5$, for example, a tunnel may be constructed without great problems concerning stability providing support is maintained close to the face. For greater values of N_s we may consider two cases:

1. For $z_0 < 100$ m (say), where a 'closed face' system of tunnelling with TBM may be adopted or, for lesser values of z_0, compressed air or face spiling might have been considered.
2. For $z_0 > 100$ m (the limit being fairly arbitrary, depending on the overall economics of the construction of a project), where the behaviour of the ground will be recognised as 'squeezing'. A different technique is then required, as described below.

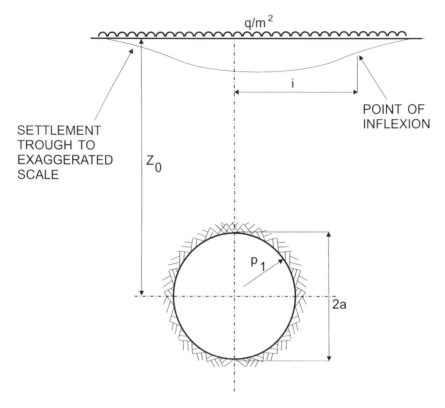

Figure 5.9 Reference diagram for stability ratio and for settlement over a tunnel.

The above elementary approach to the 'taxonomy' of tunnelling leads to a diagram such as Figure 5.10 where each zone corresponds to a different basic type of tunnelling. These limits are fairly arbitrary, depending as they do on the tunnelling technique and on the degree of variability of the ground. For instance, the presence of high water pressure in a local confined aquifer or affecting a plug of weak ground penetrated by the tunnel may suddenly set up a more severe environment.

The line $R_c = 2$ in Figure 5.10 then subdivides the continuum from the discontinuum, recognising that, between $2 < R_c < 4$ (say) the ground needs to be considered as a continuum/discontinuum. The line $N_s = 5$ establishes the approximate boundary between ground which may be stable, if excavated by open shield or equivalent if $N_s < 5$, and ground which would tend to be unstable in such circumstances, for $N_s > 5$. Soft ground techniques are then assumed by the figure as occurring for $\sigma_z < 1$ MPa. Thereafter, for $N_s > 5$, the ground will behave increasingly as squeezing

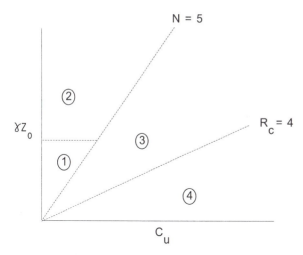

ZONE 1 : SOFT GROUND TUNNELLING
ZONE 2 : SQUEEZING ROCK
ZONE 3 : INTACT ROCK REQUIRES SUPPORT CLOSE TO FACE
ZONE 4 : INTACT ROCK SELF-SUPPORTING

Figure 5.10 Tunnelling characteristics related to strength of ground and weight of overburden.

rock with increasing values of z_0. Such a diagram should be recognised
as qualitative; in particular the critical value of N_s depends on the unsupported length of the tunnel.

The options for a tunnel in soil (soft ground) will be for support to be
provided close to the face or up to the limit of the ground supported by
a shield. Transatlantic practice has favoured the use of steel arches with
timber polings as primary support. Elsewhere the preference has been for
the use of segmental linings. Where the ground possesses real or apparent
time-dependent cohesion of a nature to allow immediate self-support of
the ground at and close to the face, a form of sprayed concrete lining
(SCL) may be adopted (see Sections 5.1.3 and 5.2.2).

Appendix 5A considers the approximation of the circular tunnel in
linearly elastic ground, Appendix 5B and 5C derive stresses and conver-
gences for ideally elasto-plastic ground around an internally supported
cylinder and sphere respectively, providing data for Figure 5.11 which
compares the two cases. This diagram emphasises the benefits conferred
by the third dimension.

The use of special expedients is described in Chapter 6. Here it merits
comment that, for a tunnel in water-bearing ground close to the surface,
immediate stability may be enhanced (and thus the opportunity provided

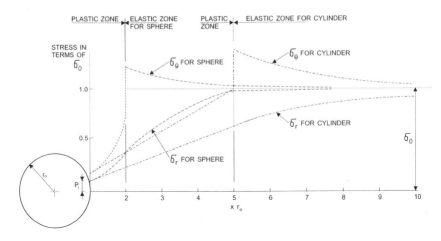

Figure 5.11 Comparison between stress pattern in elasto-plastic ground around a cylinder and a sphere (ϕ (elastic) = 30°, ϕ (plastic) = 20°, p_i = 0.1σ_0).

to adopt some type of informal support) by the use of low pressure compressed air within the tunnel, at a pressure not exceeding 1 bar (100 kPa) to avoid medical risk and time penalties. Compressed air should only be used with full compliance with regulation.

The threat from ground-water is always a factor of fundamental importance. Risk assessment should be applied to sparse data to establish the range of conditions that may fit the data. The designer should then consider the consequences. The potential rate of inflow needs to be considered in these respects:

- effect on the scheme of tunnelling;
- effect on water supply or on others dependent on the ground-water;
- effect on damage arising from water abstraction;
- practicability of stemming inflow or counteracting its consequences;
- maintenance of permanent conditions of drainage or watertightness.

Where water is present in a tunnel which is to receive a concrete lining placed *in situ*, some form of water shedding will be necessary, as described in Chapter 6, to allow satisfactory placing of the concrete. Any system of drainage to be relied upon over an extended period must be generously dimensioned and provided with means for inspection and clearance of blockages, particularly where exposed to the air, from the deposit of salts such as calcite.

The designer of the finished tunnel must know how it is to be constructed, hence the need for the total tunnel *design* process to include the design of the process of construction and the means of construction, first in conceptual terms and then developed in coordinated stages. The choice of the form of the tunnel is then a matter of optimisation, dependent on cost, time – expressed in terms of value for the Owner, and risk. Such a holistic approach permits Observational Design (Section 2.7) to be used to the best advantage.

Generally the major problems of stability arise at or close to the face of a tunnel. Exceptions occur where the ground has swelling characteristics (usually on account of the presence of gypsum, see Section 8.4) or where it may be affected by extended exposure to air or water. Design for stability must therefore be developed around possible forms or geometries of failure, related to the proposed method and means of construction. The different approaches to design are considered below, with references to more extended accounts, stressing that the expected variability of the ground qualifies the potential benefits from rigorous design methods.

Table 5.1 summarises the several potential options in approaching the conceptual design of a tunnel. Currently, for a strong rock tunnel, the options are for drill-and-blast or for advance by a Tunnel Boring Machine (TBM). The former offers flexibility in geometry, opportunity for staged construction in headings and the possibility of additional working faces. The latter offers the prospect of higher rates of advance (except in the strongest rocks) and a more even rock profile, a lesser zone of disturbed rock and reduced overbreak.

The options for a tunnel in weak rock (as defined above) are for the use of a TBM (with or without shielding, in accordance with the nature of the ground) or for the use of Informal Support (as defined in Section 5.1.3), adopting a Sprayed Concrete Lined (SCL) tunnel for a tunnel in soft ground, which may be advanced full-face or by a series of headings.

Table 5.1 Options for tunnelling

Ground type	Excavation	Support
Strong rock	Drill-and-blast or TBM	Nil or rockbolts +
Weak rock	TBM or roadheader	Rockbolts, shotcrete etc.
Squeezing rock	Roadheader	Variety of means of support depending on conditions
Overconsolidated clay	Open-face shielded TBM or roadheader	Segmental lining or shotcrete etc.
Weak clay, silty clay	EPB closed-face machine	Segmental lining
Sands, gravels	Closed-face slurry machine	Segmental lining

5.1.2 Drill-and-blast

The designer needs to take account of a number of particular features of drill-and-blast tunnelling:

- tolerance of method to rock type and to quality, extent and nature of jointing;
- tolerance to water;
- capability of providing a smooth profile;
- adaptability for variable geometry of excavation;
- extent of damage to surrounding rock;
- restrictions imposed on noise and vibration;
- risk of settlement damage.

These features are developed further in Chapter 6. At the project concept stage the possible restrictions on use of the method need to be considered, especially the question of rock damage which may control the acceptable spacing between tunnels.

Drill-and-blast tunnels may remain unlined, may be lined with *in situ* concrete or, exceptionally, with segmental linings (exceptionally in the latter respect since the degree of overbreak adds considerably to the total cost on account of the extent of annular grouting necessary). Where strong rock contains weak shear zones, segmental lining may be adopted over short lengths, possibly accompanied by a modified form of tunnel advance. Tunnelling in strong rock is unlikely to cause unacceptable ground movement provided major rock falls are prevented. Effects of draw-down of the water-table need to be considered, however, as seriously as for any other tunnelling medium (see Section 5.3 below).

5.1.3 Tunnels with Informal Support

The designer needs to take account of these special features of tunnels which adopt a form of Informal Support:

- characteristics of ground, including variability, related to design of method;
- anticipation of features which might otherwise present surprises;
- degree of tolerance to water;
- tolerance to variable geometry of excavation, including tunnel junctions;
- scheme of construction;
- time-dependence of ground behaviour;
- risk of settlement damage;
- design of intermediate stages, including special needs for face support;
- needs for observation and monitoring.

The term 'Informal Support' is used generically in this book to include systems of tunnelling which rely on primary support of the nature of bolts, dowels, anchors, mesh, arches and sprayed concrete. The system is informal, thus it is adaptable in components and pattern to the precise circumstances and it may in consequence be termed an 'informal' scheme of support which does not conform to a specific geometry. Sprayed concrete is a frequent (but not invariable) component of the system. Where arches are used, these are usually of a lattice type to avoid 'shadows' affecting the placing of the sprayed concrete (Shotcrete is a trade name widely used as a synonym for sprayed concrete). Alternatively, in squeezing ground, yielding arches may be used (Section 5.2.2). The term NATM is not used in this book for any specific type of support. NATM has been used by its proponents to describe a wide variety of tunnelling conditions so the title is confusing. Moreover, it seems advisable to avoid compromising the several forms of Informal Support with the mythology of the claims underpinning NATM, as discussed in Chapter 1.

Where Informal Support is applied to soil or weak rock, sprayed concrete will invariably be used. Such support will generally require to be applied immediately following excavation of a tunnel or of a heading to form part of a tunnel. Applications of Observational Design will therefore take a different form from that more generally applicable to Informal Support, in that the time to respond to any indication of need for supplement to the support will be limited and supplement should therefore be exceptional to the designed scheme. This does not of course inhibit the support being applied in successive layers (of shotcrete) or phases in areas which will subsequently be subjected to higher stresses, e.g. at junctions or where two tunnels are to be built close together. Support for tunnels in weak rock and soft ground is therefore differentiated by the name Sprayed Concrete Lined (SCL) tunnels, as a subset of Informal Support.

Where Informal Support is used for a tunnel advanced by a series of headings, the intermediate phases of construction assume major importance in assessing the stability of the tunnel, the face of the leading heading being possibly several tunnel (equivalent) diameters ahead of the last completed length of full lining. The design of individual headings for SCL tunnels needs particularly to take account of local deformations as a result of concentrated ground loading at sharp angles in the lining of intermediate stages (Figure 9.3) where stability will require mobilisation of local high stresses in the ground.

A common approach by the exponents of NATM to the analysis of SCL tunnels has been to adopt a 2-D ground model solved by using the Finite Elements method, with the stiffness of the ground within the area of excavation reduced in order to reproduce the observed behaviour of the tunnel in question, advanced in a particular manner. The behaviour of the tunnel is dominated by the three-dimensional ground movements in the vicinity of the face (see Figure 5.5) and hence a 2-D approach, particularly one which entails a pragmatic choice of constants to fit measurements of load or movements to suit one particular scheme of construction, has limited predictive value for any variation of the particular conditions for advancing the trial tunnel that has served for the analysis. An alternative approach is discussed in Section 5.2.2.

A feature of SCL lies in the difficulty of providing immediate support to the tunnel face. Several stratagems have been adopted in mitigation (see also Section 5.1.4):

- use of forward-inclined arches (Figure 5.12);
- use of dowels in resin-bonded glass fibre;
- use of inclined spiles driven by use of dollies to clear the periphery of the excavation;
- immediate doming of the face with the application of sprayed concrete;
- bench or temporary buttress (ground left undisturbed between side-wall footings);
- use of L.P. compressed air (Section 5.1.1);
- *sciage*, the cutting of a slot around the tunnel periphery, immediately filled with shotcrete (Bougard 1988).

A high proportion of ground loss contributing to ground movement and surface settlement is likely to be associated with inward movement of the face. Of the options described above, only the use of dowels will have direct effect on such movement, which is discussed and analysed in association with radially inward movement of the ground ahead of the face by Lunardi (1997). Vacuum well-pointing has also been suggested for the purpose but the practical problems of setting-up and dismantling would be considerable.

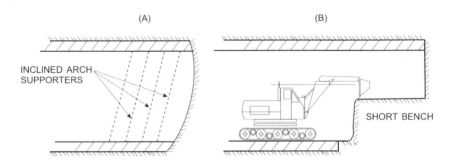

Figure 5.12 Examples of means of providing support close to the face (a) by use of inclined arch support (after Muir Wood 1993), and (b) by use of short bench.

The tunnel invert is the most vulnerable part of the SCL tunnel lining, on account of adhering spoil and rebound from sprayed concrete, possible damage under construction traffic and the demanding requirements for forming satisfactory joints in the lining where headings have preceded the full section. Furthermore, it is often concealed from view soon after construction. One precaution in design is to avoid the temptation prompted by economy in excavation for too 'flat' an invert by maintaining a radius no greater than, say, the mean diameter of the tunnel, the invert then being partially backfilled with tunnel spoil ('clay running') to provide a surface on which to run plant.

It is likely that bending stresses in sprayed concrete tend to be overestimated by theoretical calculations which do not consider the plastic behaviour of the concrete in the early hours of its life when much of the development of load expected to contribute to bending stresses will be occurring.

5.1.4 Squeezing ground

Squeezing ground represents a particular form of rock which can only practically be penetrated by adherence to a stratagem which takes full cognisance of the potential problems of excessive encroachment of the ground into the tunnel. Kovári (1998) illustrates the problem to be solved in that, unless supported by unacceptably heavy support, the ground will invade much of the cross-section of the tunnel before – providing collapse is prevented – equilibrium may be attained. To cope with this threat, a yielding system of support is needed, possibly based on specifically designed rock-bolts, possibly on yielding arches coupled with spiles, capable of tolerating high local strains. Sprayed concrete will tolerate only limited compressive strain, which may be as low as 0.2%, prior to spalling. The Alpine tunnels have demonstrated squeezing ground at its most intractable, where every expedient of face support, inclined spiles 'slotted' shotcrete and yielding arches have been successfully adopted. Kovári (1998) illustrates how, once the ground for a pilot tunnel has been stabilised, trimming back to the line of the full section will need to be associated with a high level of support. Once the extent of convergence can be estimated, an alternative approach will entail oversize excavation to the full section with the accepted convergence respecting the dimensions of the minimum section (see also Section 5.2.2).

5.1.5 Tunnels driven by TBM or Shield

The term 'Tunnel Boring Machine' (TBM) is customarily used to describe a full-face machine used for advancing tunnels in rock, usually but not necessarily making a circular cut. A shield implies a device to provide immediate support to the ground (possibly relieved by the slight projection, the relief or bead – the latter possibly applied as a run of weld metal – of the cutting edge) and thus that it is used in weak ground. A shield may be fitted with a mechanical rotary cutter or it may serve as an essentially cylindrical protection, within which excavation is undertaken by hand or by use of a separate machine (back-actor, road-header or similar). In this book, for simplicity, the term TBM designates a full-face shielded or unshielded machine, while a shield is confined to a self-propelling device for providing ground support around the excavation.

TBMs may then be classified in the following working modes:

- unshielded TBM, where provision is made between face and tail of the machine to protect operatives and equipment against rock falls but where no continuous support is provided by the machine;
- shielded TBM, where ground support is provided by the machine itself, possibly using weak bentonite/cement grout outside the TBM for the purpose in weak rock where steering of the machine requires some over-cutting of the ground.

A shielded TBM may then be further classified as a machine with no additional means for face support, other than the possible use of shutters between the cutters, or as a 'closed-face' machine, depending on some form of balancing of the pressure of the ground and of ground-water. Where support is required along the TBM, a drill may operate, possibly through slots in the skin, so that bolts may be provided close to the face. A closed-face machine may rely on maintaining pressure at the face by means of a compressed air reservoir, using a slurry medium from the ground itself where suitable or by additives to the spoil as bentonite, foaming agents or other means to permit pressure to be transmitted to the ground (rather than to the ground-water). An alternative means is the Earth Pressure Balancing Machine (EPBM) whereby the pressure at the face depends upon the coordination of the advance of the machine with the ejection of spoil by an inclined rotating Archimedes screw conveyor. Machines of these types may be designed to operate in open or closed mode, the former allowing more rapid progress in suitable ground.

For maximum rate of progress and to assist steering, the machine may have an articulating peripheral joint such that the forward section with cutter-head may advance separately from the rear to the extent permitted by the stroke of the intermediate thrust rams.

The profile of a shield- or TBM-driven tunnel will normally be circular (although elliptical shields have been designed and figure-of-eight shields and even more complex configurations have been used in Japan) with the lining following close behind or within the tail of the shield. Where the lining is built behind the tail of the shield, broken ground may require intermediate support in the crown to bridge this gap, probably by trailing bars or plates supported on the last erected ring of the lining, an expedient adopted by Brunel for the Thames Tunnel (see Chapter 1). For a closed-face machine, design of the seal between machine and lining, built within the tail, is a critical feature. Rubber seals have now been generally superseded by grease-impregnated steel wire brush seals.

Where the ground is variable, selection of the appropriate type of TBM (see also Section 6.3) must take account of the differential efficiencies of each type in the expected mix of ground types, having regard to the tolerance or adaptability of each if expectation is exceeded in any respect, and the special expedients that may be adopted for the adaptable forms of machine.

5.2 Design of the support system

5.2.1 Steel arches

Steel arches from rolled sections continue to be used for support, mainly where loose rock causes any 'active' system of support to be too difficult to install. The design of passive support by steel arches is based on the notion of the unstable rock wedge in the crown, or possibly asymmetrically, to be supported, with the arch in its turn buttressed against the rock around the remainder of the periphery of the tunnel, to limit bending stresses.

The design of foot-blocks is vital to the success of the system of support, in relation to the bearing capacity of the ground, reduced as it may be by local disturbance or excavation. Where a tunnel is to be advanced by crown heading and bench, temporary foot-blocks are needed at bench level, with provisions for subsequent extension legs (Terzaghi 1961). The general approach advocated by Terzaghi, similar to that advocated by Kommerell (Figure 1.2) remains valid to the present day.

The weakness of steel arch support concerns the load at which failure may occur by lateral buckling and torsion. Their load-bearing capacity may be increased several-fold by providing continuous bedding against the rock in place of the traditional timber packing. The principal of a means for achieving this objective by the inserting between arch and rock a bolster made in porous fabric filled with a weak sand/cement grout has been described by Muir Wood (1987), illustrated by Figure 5.13, and an application described by Craig (1979). Experiments commissioned by the National Coal Board demonstrated the considerable increase in bearing capacity by this means. Theoretically, making simplifying assumptions, the load-bearing capacity of the bag per unit length of arch, for a non-cohesive, no-tension, filling, may be calculated as:

$$\text{Load} = 2t[\exp(2K_p b \tan \phi/d) - 1]/\tan \phi \qquad (5.9)$$

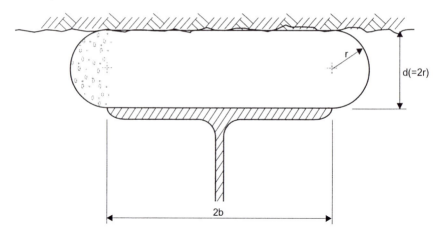

Figure 5.13 Reference diagram for bagged packing.

where t = tensile strength of bag/unit length, with b and d as indicated on Figure 5.13, and $K_p = (1 + \sin \phi)/(1 - \sin \phi)$ where ϕ relates to filling material.

Toussaint–Heinzmann yielding arches have a bell-shaped section, sections being coupled by friction clamps, as illustrated by Széchy (1970). Kovári (1998) describes their use in squeezing ground, with thrust capacities for an arch varying between 300 and 600 kN. The coupling needs to occur around a part of the perimeter of constant curvature.

5.2.2 Informal Support for tunnels in weak and squeezing ground

There is something of a paradox in the fact that the most critical conditions for the stability of the ground around a tunnel occur locally to the face. Yet, if the ground is stable in the three-dimensional stress system which occurs in this vicinity, installed support at first bears little load. This share increases as the support may be loaded by convergence of the ground but principally by advance of the face which throws the load from the ground 'dome' onto the last completed section of support (Figure 5.6). A preliminary approach may assume an elastic ground reaction, with tests for sensitivity across wide variation, and appropriate for immediate and for long-term loading. A simple basis for the coefficient of ground reaction proposed by Muir Wood (1975a) is provided in Appendix 5A.

As described in Section 5.1.1, the stability of the ground in the vicinity of the face of a tunnel depends on the formation of natural doming (the 3-D equivalent of arching) of the ground. The time-dependent analysis of the ground in full 3-D is an elaborate approach only justified in particularly delicate circumstances (see for example Higgins *et al.* 1996 concerning tunnelling for the Jubilee Line Extension in Westminster, London near to the tower of Big Ben). An alternative approach, which has similarities to that proposed by Renato and Karel (1998) for squeezing ground, for more general application is outlined below:

1. The physical properties of the ground are introduced as constitutive equations in appropriate detail through CRISP or another program.
2. The heading (or tunnel) is represented as a circular cylinder of corresponding cross-sectional area located centrally to the heading.
3. The area around the face of the representative cylinder is analysed by axi-symmetric 3-D finite elements at times after excavation corresponding to the rate of advance, i.e. the period over which the ground remains unsupported for this affected length of heading.
4. For the supported section of the heading, analyses would be undertaken in 2-D, taking account of transmitted loading from 3 above, and for the period prior to full lining or prior to opening up of a heading.

Lunardi (1997) offers an alternative Lower Bound approach which takes particular account of the stability of the core of ground ahead of the face, the 'advance core', within the cylindrical projection of the tunnel. This is an interesting approach, taking proper cognizance of the significance of the stability of the stressed dome of ground around the face, itself supported on the advance core, a natural feature of any form of tunnelling, as the most vital element of the design of the interactive support. Lunardi (1998) describes application of the ADECO-RS (Controlled DEformation in Rocks and Soils) method to the design of tunnels in squeezing ground on the High Speed Rail Link between Bologna and Florence.

The debate continues between the virtues of dry-process and wet-process shotcrete (the water for the former being added at the nozzle), the latter claiming improved control and less rebound, the former economy in cost (International Tunnelling Association 1991). The performance of shotcrete may be improved by the addition of fibre reinforcement (Kasten 1997). Steel fibre of different geometries has usually comprised fibres about 50 mm long and 0.5 mm diameter or equivalent.

Where rock-bolts serve to create an arch of self-supporting rock around a tunnel, a simple basis for analysis is that proposed by Lang (1961), described in Appendix 5D. In weak rock, where convergence as the tunnel advances may be found to cause excessive stresses so as to damage the anchorage of the bolt, an expedient (Kidd 1976) is to tension the bolt to close joints that may have opened with the relaxation of the excavation, then to release much of the tension prior to further excavation.

Water may be permitted to flow freely behind the sprayed concrete lining and only stopped off (Section 6.3) when the secondary concrete lining is in place which, in these circumstances, has therefore to be designed to support the water load together with rock and any other loading relating to full or partial failure or deterioration of the primary support. While anchor systems may be designed to have an extended life and thus to be treated as part of the permanent work, the elaborate methods of protection that are then necessary entail an additional cost only acceptable for major anchorage systems for large caverns and not normally economic for steel supports as bolts, dowels or mesh.

5.2.3 Segmental linings

Segmental linings are loaded by the ground, through the agency of annular grouting, by expansion of a ring against the ground or by movement of the ground onto the lining at some distance from the tunnel face. Generally therefore the ground loading of such linings may be assumed to occur in a two-dimensional manner. This is often based on the fiction that the tunnel is tightly inserted into the ground which is only then permitted to relax to find a stable state between the perforated ground and the tunnel lining. A simple approach is described by Muir Wood (1975a) and extended by Curtis (1976), whereby the stresses in the lining resulting from the combination of hoop loading and bending may be calculated in terms of assumed linear elastic characteristics of the ground and the lining (Appendix 5A).

The above approach, combined with a sensitivity analysis with the critical parameters varied between limits, may well provide an adequate basis for design of stiff or articulating linings in simple conditions of ground loading where depth of overburden z_0 (Figure 5.9) is at least $2a$, where a is the effective tunnel radius. For lower ratios of z_0/a, critical failure modes should be considered in order to provide a Lower Bound solution.

The stiffness in bending of the lining in the plane of the ring is small in relation to that of the ground, except in very weak and readily deformable soils such as unconsolidated silts, or where, from consideration of local uneven loading or low depth of overburden, the lining is deliberately stiffened. For this latter purpose, joints between adjacent rings may be staggered, with stiffness increased by ensuring tight contact across the joints between adjacent rings, and the longitudinal joints between segments deliberately strengthened as described by Craig and Muir Wood (1978) for the construction of Kings Cross Underground Station on London's Victoria Line.

There is great virtue in the use of unreinforced concrete for tunnel linings, recognising the dominant influence of static loading and the avoidance thereby of concern for long-term corrosion. Expanded linings

(Figure 5.14) usually combine the virtue of low stiffness with practical considerations of erection and stressing by deliberately permitting free rotation at the longitudinal joints.

The most critical potential hazard for the Cargo Tunnel at Heathrow Airport (Muir Wood and Gibb 1971), eliminated so far as practicable by research of the prior use of the site and by ground investigation, was the possible inadequacy of clay cover. Under Runways 5 and 6, the inaccessible parts of the airport, which were of course the most vital to the operation of aircraft, there remained the possibility of a dip in the surface of the London clay. This might have originated from local erosion prior to the deposit of the overlying Taplow gravels or it might have resulted from local over-excavation by gravel dredgers into the upper surface of the London clay in which the 10.3 m diameter tunnel was being constructed under a total cover of less than 8 m (Figure 5.14). As a result of consideration of possible failure modes, an acceptable criterion was for at least 1.1 m cover of London clay above the tunnel beneath the water-bearing gravels. Any depression below this level was to be detected by overlapping probe holes inclined upwards ahead of the tunnel face, the contingency plan entailing a scheme for injecting the ground to make good any deficiency in the requisite clay cover. In the event, no such contingency was drawn upon.

Expanded linings are not only simple in construction, eliminating the need for fixings or for grouting for structural reasons, but may contribute to reduced ground movements by comparison with grouted linings or Informal Support. Where the joint stiffness is less than that of the general cross-section of the ring, the maximum bending moment that may be developed within a segment may be approximately calculated by consideration of the relative stiffness of the joint (see Appendix 5A).

Tunnels are normally built for long life, with a consequential need to consider the effects of the options of abandonment and collapse. For expanded linings, no fastenings are required between segments and the dimensions of a segment may be chosen to provide an 'aspect ratio' sturdy enough to survive accidental damage during the processes of handling and erection. Section 1.5 describes briefly the evolution of the expanded lining. The choice of the number of segments to the ring and the preferred width of ring affects the scheme for expanding the ring. The simple expedient is to use tapered segments or wedges. These are replaced by jacks where ring width is small in relation to the diameter of the tunnel (c.f. Figures 1.3 and 5.14). The Don-Seg lining, with every segment tapered, provided greater tolerance to slight variation in excavated diameter than wedge-block linings with reduced numbers of wedged segments (Clark et al. 1969).

Where segments are to be reinforced, in the presence of aggressive ground-water or an internal source of corrosion, the several options include electrical bonding of reinforcement (which may have other desirable

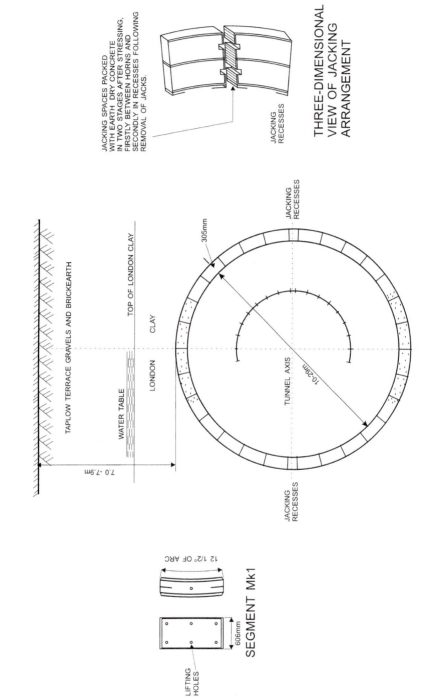

THREE-DIMENSIONAL VIEW OF JACKING ARRANGEMENT

JACKING SPACES PACKED WITH EARTH DRY CONCRETE IN TWO STAGES AFTER STRESSING, FIRSTLY BETWEEN HORNS AND SECONDLY IN RECESSES FOLLOWING REMOVAL OF JACKS.

JACKING RECESSES

TAPLOW TERRACE GRAVELS AND BRICKEARTH

TOP OF LONDON CLAY

CLAY

LONDON

WATER TABLE

7.0 - 7.9m

JACKING RECESSES

305mm

10.29m

TUNNEL AXIS

JACKING RECESSES

12 1/2° OF ARC

606mm

SEGMENT Mk1

LIFTING HOLES

Figure 5.14 Heathrow Cargo Tunnel (after Muir Wood and Gibb 1971).

consequences relating to the use of the tunnel, such as the security of railway signalling), protective (epoxy resin) coating to the reinforcement, and high density concrete with high resistance to percolation by water and to diffusion of ions of H or Cl. External coating of the segments is liable to be damaged during handling. An alternative, used particularly for expanded linings, is the use of steel or glass fibre reinforcement, taking due account of the brittleness and time dependence of liability to alkaline corrosion of the latter.

Where an expanded lining is impractical, e.g. where the nature of the ground requires the lining to be built within the tail of a shield or where the peripheral surface of the exposed ground is too uneven to provide a satisfactory bearing surface, the use of the tunnel largely determines the details of the optimal bolted form of segmental lining.

Flanged segments continue to be used with connecting bolts between longitudinal and circumferential joints, particularly for small tunnels advanced by hand or with simple minor items of plant. The trend has however been towards the increased adoption of reinforced concrete segments with a flush intrados, apart from recesses for fixings, except for those circumstances where exposed bolts also serve for other attachments or where a secondary lining is required, for example protective brickwork for sewer tunnels. Attachment between flush segments adopts one of several arrangements, of which the most widely used are either bolts formed to an arc of a circle or bolts slightly inclined in relation to the surface of the segment which engages with a socket cast into the adjacent segment (Figure 5.15).

For the Channel Tunnel, even bedding against the chalk marl of an expanded lining was provided by the use of pads for the initial state, with grouting to follow (Figure 5.16), for the British section. For the French section, where the lining was required to withstand external water pressure, a bolted, gasketted lining was used (Figure 5.17).

While new tunnels are unlikely, except in remote areas of developing countries, to be constructed in masonry or structural brickwork, the engineer may well be required to analyse the adequacy of an existing lining. Codes of Practice for brickwork tend to be unnecessarily prescriptive for the analysis of existing tunnel linings. Too often resort is made by the inexperienced engineer to a computer program for an elastic continuum, which will then indicate local overstressing. The appropriate approach is to make use of the notion of the 'thrust line', described in Appendix 5E. This approach may well serve as a first check also on a concrete lining, where transverse tension cracking in bending is acceptable for an ultimate state.

Sealing against the entry of water into a bolted lining is commonly achieved by girdling each segment with a shaped polymeric gasket, compressed against a twin gasket by tightening the bolts. Such seals have been

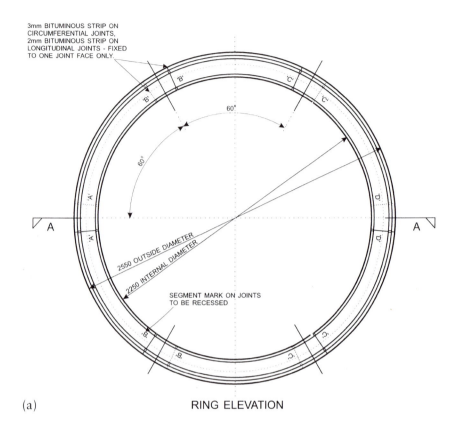

3mm BITUMINOUS STRIP ON
CIRCUMFERENTIAL JOINTS,
2mm BITUMINOUS STRIP ON
LONGITUDINAL JOINTS - FIXED
TO ONE JOINT FACE ONLY.

'B'

'B'

'C'

'C'

60°

60°

A

A

'A'

'A'

'D'

'D'

2550 OUTSIDE DIAMETER

2250 INTERNAL DIAMETER

SEGMENT MARK ON JOINTS
TO BE RECESSED

'B'

'B'

'C'

'C'

(a) RING ELEVATION

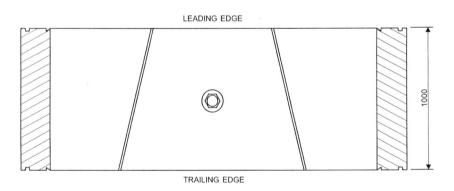

LEADING EDGE

TRAILING EDGE

1000

(b) SECTION A-A

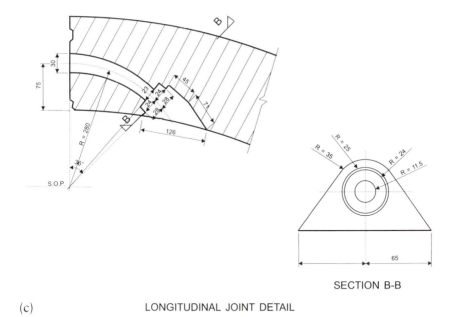

SECTION B-B

(c)   LONGITUDINAL JOINT DETAIL

Figure 5.15 Typical expanded lining with bolted trapezoidal segments. (a) Ring elevation.
(b) Section A–A. (c) Longitudinal joint detail. (All dimensions in millimetres)

tested against pressures up to 16 bar for the Storebaelt Tunnel (Elliott *et al.* 1996). Alternative methods use polymeric seals of a type which expand on contact with water. There are also many injected sealants, usually used as a second line of protection in association with a gasketted seal.

Annular grouting is achieved by cement grout with additives as appropriate for control of setting time and to avoid washout by flowing water. Provision should always be made to permit grouting close to the last erected ring, with the process only relaxed as a result of deliberate design or experience.

The principal features of pipe-jacking for the construction of pipelines, headings and tunnels of a variety of shapes concern their construction, are described in Chapter 6. Structural design has to consider loading during construction, especially non-uniform longitudinal thrust, also uneven loading in adjustments to direction of thrust. Following completion, loading will correspond to that on a stiff formal tunnel lining rather than the Marston/Spangler type formulae for pipes buried in trench (Bulson 1985).

5.2.4 Tunnel junctions and enlargements

The design of junctions in tunnels or of sudden changes in section of the tunnel receives too little attention (Muir Wood 1970), particularly in

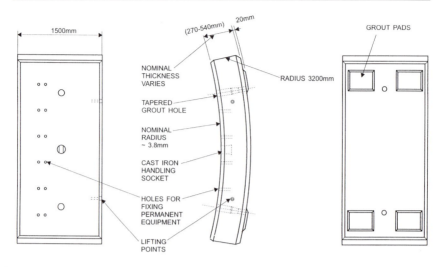

Figure 5.16 Typical precast concrete segment for UK section of the Channel Tunnel (after Eves and Curtis 1992).

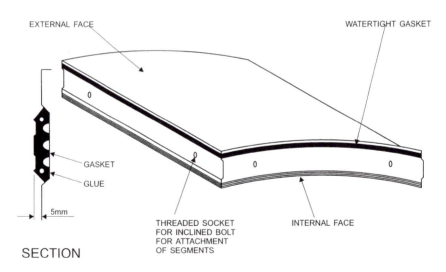

Figure 5.17 Typical precast concrete segment for French section of Channel Tunnel (after Barthes *et al.* 1994).

relation to intermediate phases of construction. The guiding rules are that oblique junctions, where unavoidable, should be achieved by means of an expansion to contain the two tunnels, on the pattern of the traditional step-plate junction for segmental linings. Square junctions should entail a minor tunnel junctioned into a major (continuing) tunnel since the junctioning of two tunnels of much the same size presents considerable problems of temporary support during construction. The problem may be averted by including an enlargement, costly for a shield-driven tunnel, in the continuing tunnel in the vicinity of the junction. The objective should be to include in the continuing tunnel special segments which may later provide the frame to the opening for the junction, and temporary segments later removed to form the opening. In this way, there is no major disruption to the construction of the continuing tunnel (Figure 5.18).

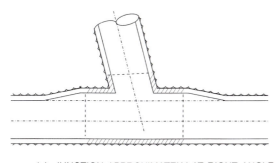

(a) JUNCTION APPROXIMATELY AT RIGHT ANGLE

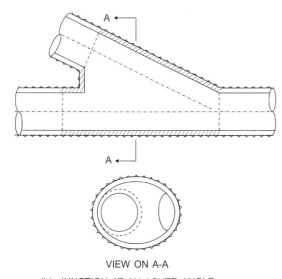

VIEW ON A-A

(b) JUNCTION AT AN ACUTE ANGLE

Figure 5.18 Junction of SCL tunnels of similar size.

5.3 Ground movements and surface settlement

The anticipation of the possibility of ground movements associated with tunnelling, leading to damage to structures or services, is a vital feature of overall tunnel design. Ground movements may be considered in two parts:

1. immediate ground movements caused by loss of ground, or changes in ground stresses, in the vicinity of the tunnel face;
2. long-term movements generally associated with ground consolidation caused by changes in effective ground stress, which may be associated with flow of water towards the tunnel (or possibly swelling, for a water or sewer tunnel, as a result of leakage from the tunnel).

Loss of ground at a localised region in a soil at depth z_0 below the ground surface gives rise to axisymmetric strains which form a cone of depression at ground surface. Where there is a continuous line of such punctuated ground loss, the superimposed 'cones' create a settlement trough of depression (Figure 5.9), with a shape often assumed as that of a normal distribution curve which then permits simple relationships between parameters of the curve. Thus, where w, the depression, attains a maximum value w_{max} with points of inflexion occurring at a distance i from the centre-line, the area of the trough $V_s = 2.5\,i\,w_{max}$. With such a simple model, the ratio of i/z_0 for a tunnel of diameter $2a$ with axis at depth z_0 was related by Peck (1969a) on the basis mostly of hand-driven tunnels in the USA as $i = 0.2(2a + z_0)$. For cohesive soils the trough is certainly wider than the value given by this formula and several authors have looked for more reliable relationships, undoubtedly affected by many different factors that no unique relationship can express. The formulae suggested by O'Reilly and New (1982) were based on analysis of a number of UK tunnels:

$$i = 0.43(z_0 - z) + 1.1 \text{ m for cohesive soils}$$

$$i = 0.28(z_0 - z) - 0.1 \text{ m for granular soils}$$

within stated limitations, thus indicating the value of i at depths z above the tunnel and up to the surface. Mair *et al.* (1993) suggest, on the basis of further data, that a reasonably good fit for tunnels in clay is given by:

$$i = Kz_0, \text{ where } K = 0.175\lambda + 0.325 \text{ and } \lambda = z_0/(z_0 - z)$$

For a granular soil, the phenomenon is visualised as intergranular movement under gravitational force. The relationship between V_t, ground loss into the tunnel measured as m^3/m and V_s will depend on the initial state of the soil: $V_t/V_s > 1$ for a dense soil, $V_t/V_s < 1$ for a loose soil. For very loose soils, disturbance by tunnelling has on occasion caused factors well in excess of unity. For a natural soil, this risk may be caused by base exchange since deposition, for example, giving rise to sensitive soils (e.g. 'quick' clays) or by slight cementation between loosely packed grains (collapsing soils). For man-made deposits, the procedure of deposition may be the dominant factor; attempts to construct a jacked heading through an end tipped chalk railway embankment caused gross settlement as disturbance caused finer material to flow into cavities between the large boulders at the foot of the embankment.

For a clay, the volume per unit length of the settlement trough will be determined in relation to stress redistribution in the vicinity of the tunnel, with losses into the tunnel V_t causing stress relaxation.

From the development of such simple initial concepts (Attewell *et al.* 1986) settlement may be estimated for particular circumstances, taking account of comparative experience elsewhere. The first consideration is that of the ratio $p(= V_t/A)$, where A represents cross-sectional area of the tunnel.

For a shield-driven tunnel, the contributory factors to V_t may be subdivided as (Muir Wood 1970):

- loss at face;
- loss around shield;
- loss behind shield.

This indicates where measures may be taken to minimise ground movement:

- face loss – by face support (e.g. by face rams, by GRP spiles or by earth pressure balancing);
- losses around shield – shield alignment, minimum bead, short shield, use of weak grout for TBM;
- loss behind shield – immediate stressing of expanded lining.

Particular attention was given to each of these factors for the Heathrow Cargo Tunnel (Muir Wood and Gibb 1971), resulting in a value of $p = 0.3\%$. Mair *et al.* (1993) provide a fairly coherent pattern of 'green field' settlement of tunnels in London clay related to contributory factors, including the value of N_s.

For a SCL tunnel, the face is unsupported and, while the support to the ground is provided in a manner to permit direct loading by the ground, settlement is greater than for a comparable well-managed, shield-driven tunnel (Simîc and Gittoes 1996, Bowers *et al.* 1996), notwithstanding claims which have been made to the contrary. A multiple drift SCL tunnel entails much redistribution of stresses in the ground between each phase of advance, with further expected contributions to settlement although the procedure should cause reduced loss of ground from an incoming face than for a full-face tunnel with unsupported face.

Settlement calculations are essential where buildings or services exist in the vicinity of a projected tunnel. The existence of stiff structures may considerably modify the calculated 'green field' settlement trough and this feature must be incorporated in the calculations. The extent of modification depends on the stiffness of the structure against compression and tension, also the stiffness in bending (Boscardin and Cording 1989, Potts and Addenbrooke 1996). Making allowance for such modification, it is then possible to examine the risk of causing structural damage (Mair *et al.* 1996), making use of the criteria for unacceptable damage derived by Burland and Wroth (1974). The critical effect of settlement does not lie in its direct magnitude (apart from piled foundations) but arises from the associated curvature and consequential compressional and tensional soil strains. Long-term settlement, attributable to consolidation, is often found to contribute relatively little to structural damage since it is generally less in magnitude and extends more widely than the immediate settlement. Piped services may tolerate a certain degree of curvature, reflected for stiff pipes as angular deflection at the joints, but this tolerance depends on the material of the pipe and the nature of the design of the joint (Attewell *et al.* 1986, Bracegirdle *et al.* 1996).

Ground movements associated with pipe-jacking follow a similar approach to that described above, with particular modifications. First, in very weak soils at shallow depth, there is a greater risk of heave caused by excessive thrust. Second, as discussed in Chapter 6, drag between the tunnel (particularly for a rectangular heading) and the overlying soil may, unless controlled, lead to irregular ridges in the soil surface with consequences to buried services.

Where two or more tunnels are constructed sufficiently close for interference between the patterns of ground movement caused by each, settlement is usually found to be greater than that estimated by direct superimposition, doubtless as a result of non-linear effects of incremental variations of the stress tensor.

The effects of piezometric change need to be taken into account in calculating long-term settlements associated with consolidation (Appendix 5G). Relatively slight inflows of water into a tunnel have been the cause of serious ground movements affecting structures as described in Chapter 8. Rock tunnels may not be immune from causing unacceptable ground movements where water inflow may, by way of the jointing system, cause draw-down in overlying fine sediments See, for example, the account of the effect from driving the Hong Kong Bank seawater tunnel (Troughton *et al.* 1991).

Where ground movements would otherwise be excessive, control may be achievable by one or other expedient. Ground movement caused by losses at the face may be controlled by spiling or dowelling in such a manner as not to obstruct excavation (Section 5.2.2). Some form of consolidation or jet-grouting as an umbrella above the plug of ground in the tunnel face may provide an alternative solution (Chapter 6). There are several means of providing a hood above a tunnel through an unstable area of ground. Most commonly, this procedure has been adopted to penetrate loose and weathered ground adjacent to a tunnel portal. The ground may be consolidated as an arch formed from a series of jet-grouted lengths of soil or rock fragments of approximately cylindrical form. Alternatively, crown-bars may be driven at intervals in the form of an arch; these crown-bars may be formed from steel drill-hole casings subsequently filled with concrete. The particular solution needs to suit the ground, the access for construction and the available resources. If the ground is fine and water-bearing, the tubular crown-bars may need to be closely spaced with the tubes linked by external clutches, or a form of overlapping 'secant' piles may be adopted. In other circumstances, the spacing between bars will allow arching of the soil to occur and for the spaces to be filled or grouted as a top heading is advanced. The form of construction will need to ensure that adequate support to the crown-bars is provided as part of the sequence of construction as the tunnel advances, including consideration of support at the face to the leading end of each bar.

Grouting to control ground movement may take the form of permeation grouting, to prevent loosely packed material from collapsing or jet-grouting to form a slab above a tunnel in weak ground. In weak clay, jet-grouting may cause ground heave, a phenomenon experienced on a considerable scale in Singapore for the Mass Rapid Transit System in 1985 (Marchini 1990).

Compensation grouting entails the forcing of grout into the ground between, or ahead of, the tunnel and the structure to be protected. The procedure is normally to sink shafts appropriately positioned in relation to the zone of ground to be treated (Shirlaw 1996). Grout-holes are drilled radially from the shaft, and provided with means for sleeve grouting (*tubes-à-manchette*, Figure 1.5). There follows a process of preconditioning, whereby the ground is locally consolidated and possibly fractured so that further injection will cause immediate ground response; the compensation of ground movement or settlement is achieved by injecting, in controlled volumes, a bentonite-cement grout of suitable properties at controlled pressures at locations indicated by a 'real time' control program which combines records of the grouting system with movements at selected observation points. The choice of level of the compensation grouting will be partly dictated by the geology, partly to ensure that ground settlement affecting buildings and services may be limited without imposing excessive loading on the advancing face of the incomplete tunnel. Where a tunnel of radius a is situated at depth z_0, compensation grouting at depth h may, as a first approximation, be considered as influencing the tunnel loading by a factor proportional to $\{h/[z_0 - (a + h)]\}$. The essence of the design, which must be of an observational nature (Section 2.7), is to ensure a predesigned scheme for correcting unexpected departures from prognosis in relation to limits of acceptable ground movements and to loading on the tunnel, adopting a system of triggers to indicate the need for special measures. Ideally, the grouting needs to be synchronised with the occurrence of the potential settlement movement to minimise the pressures required from the grouting operation (see also Chapter 9).

5.4 Pressure tunnels

Generally, the design of a tunnel support system is dominated by the requirement to support external ground and, often, water loading. Pressure tunnels, where internal pressure may exceed reliable minimum ground loading, present a special case in which the tunnel needs also to be designed against internal pressure. For this reason, pressure tunnels often incorporate a steel lining which may itself require, in the presence of high external water pressure, to be anchored into the surrounding concrete (Jaeger 1955, Kastner 1971). To summarise much written on the subject: (1) the tunnel may be designed as a concrete-lined tunnel which remains intact by virtue of the support of the surrounding ground or (2) full internal pressure is contained within a steel lining, or (3) where the concrete lining and the surrounding rock, possibly stiffened by grouting, provide reliable contribution to containment, by a composite 'thick-walled pipe' solution. The problem with any intermediate solution of type (3) concerns the relative stiffness of rock and lining which are required to be compatible in any calculation of load sharing. Circumferentially prestressed linings (Kastner 1971) have also been used for pressure tunnels.

Traditionally, calculations based on a concrete lining in rock have calculated seepage forces as if the lining is cracked and has an effective permeability no less that that of the rock. The critical failure mode considers the dislodging of a wedge of ground from above the tunnel. This is usually a safe assumption but Lu and Wrobel (1997) suggest that it may be unnecessarily conservative, where the rock permeability is known and where calculation establishes that cracking of the lining under internal pressure results in the effective permeability of the reinforced concrete lining remaining an order of magnitude less than that of the ground.

Where a steel lining is provided, one fundamental feature of the design concerns the prevention of buckling as a consequence of rapid draw-down of a slightly leaky tunnel contributing to high external water pressure.

Jaeger (1955) and Kastner (1971) describe several examples of failure of pressure tunnels with accompanying explanations of the deficiencies in design and construction.

5.5 Aids to design calculation

The Appendices which follow illustrate a number of simple approaches to different aspects of tunnel design which the Author has found useful in practical application, and which are rarely found in juxtaposition. These may permit simple initial estimates, frequently all that the uncertainty of data may justify.

Appendix 5A The circular tunnel in elastic ground

'Neither the ground nor, usually, the tunnel lining, behaves in an elastic manner' (Muir Wood 1975a). There remains merit nevertheless in the ability to establish a closed form of solution for the elastic case, as a first indicator of stresses and deformations, particularly in investigating sensitivities to variation in the moduli of ground and lining. For an 'elliptical' mode of deformation, a solution starts (Muir Wood 1975a) from the Airy stress function in polar coordinates (Figure 5A.1):

$$\phi = (ar^2 + br^4 + cr^{-2} + d)\cos2\theta$$

where

$$\sigma_r = 1/r\partial\phi/\partial r + 1/r^2\partial^2\phi/\partial r^2, \quad \sigma_\theta = \partial^2\phi/\partial r^2 \text{ and } \tau_{r\theta} = -\partial/\partial r(1/r\partial\phi/\partial\theta)$$

solved for conditions of plane strain, so that

$$\epsilon_z = 1/E\{\sigma_z - \nu(\sigma_\theta + \sigma_r)\} \text{ etc.,}$$

whence radial deflection, u_0, may be expressed in terms of σ_r and σ_θ.

If we suppose a thin lining, the maximum bending moment in the lining induced by a displacement $u_0 = \hat{u}_0 \cos2\theta$ is:

$$M_{max} = \pm 3\hat{u}_0 E_l I r_0^2 \tag{5A.1}$$

By equating u_0 for the ground and for the lining, the equation for M_{max} may be expressed (Curtis 1976) as

$$M_{max} = \frac{p_0 r_0^2}{4 + (3 - 2\nu)\, Er_0^3/[3(1 + \nu)(3 - 4\nu)\, E_l I_l]} \tag{5A.2}$$

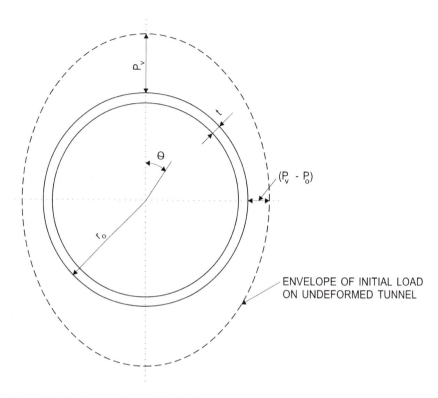

P_v

θ

$(P_v - P_o)$

r_o

ENVELOPE OF INITIAL LOAD
ON UNDEFORMED TUNNEL

Figure 5A.1 Reference diagram for loading on a circular tunnel.

(where suffix l relates to the lining) for full bond between the ground and the lining, or as:

$$M_{max} = \frac{p_0\, r_0^2}{(10 - 12v)/(3 - 4v) + 2\, Er_0^3/[3(1 + v)(3 - 4v)\, E_l I_l]} \quad (5A.3)$$

where slip occurs between the ground and the lining.

In a similar fashion, expressions for $\hat{u}$ and the mean circumferential load in the lining, N, may be derived (Muir Wood 1975a). These quantities have been plotted by Duddeck and Erdmann (1982) for the assumptions of $v = 0.3 M_{max}$ and N may be derived from Figure 5A.2 for the case of full bond between ground and lining. The maximum lining stress may be estimated by adding mean hoop stress to the maximum bending stress.

The direct loading in the lining may be reduced, in appropriate circumstances, by high circumferential compressibility. The bending stiffness will

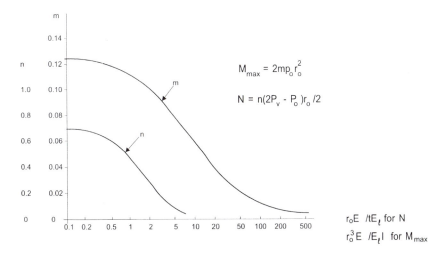

$$M_{max} = 2mp_or_o^2$$

$$N = n(2P_v - P_o)r_o/2$$

$r_oE\ /tE_t$ for N

$r_o^3E\ /E_tI$ for M_{max}

Figure 5A.2 Hoop force and maximum bending moment in continuous circular lining with full bond between lining and ground ($v = 0.3$).

then be low and the reduction in N, ΔN, may be estimated through the relationship:

$\Delta N = \sigma_0r_0/(1 + R_s)$, where the compressibility factor, R_s, may be written as:

$R_s = r_0E/tE_l(1 + v)$ for a thin lining.

The coefficient of ground reaction, λ, may be derived as:

$$\lambda = 3(3 - 2v)E/[4(1 + v)(3 - 4v)r_0] \tag{5A.4}$$

or, for $v = 0.3$, $\lambda = 0.76E/r_0$, which may be applied as a first estimate of the effect of small distortions of a circular or non-circular lining.

Muir Wood (1975a) discusses the effect of joints of lower stiffness than the remainder of the lining, where the joint stiffness is designated as I_j. This is an artificial concept since the joint has normally negligible circumferential length. An alternative approach is to consider the rotation, θ, at the joint when a unit bending moment is applied. For the lining between joints, the rotation for a unit circumferential length subjected to a unit bending moment is $1/E_lI_l$. For a number of segments, n, in a ring, where, say, $n \geqslant 8$, the effect of rotation at the joints may be approximated by reducing the value of M to be sustained by the ring by a factor $1/(1 + \theta E_lI_ln/2\pi r_0)$, with stiffness reduced and deflections increased correspondingly.

Appendix 5B Cylindrical cavity with internal support

A classical problem is that of a hole in a homogeneous uniformly stressed solid which behaves linearly elastically up to a certain state of stress and perfectly plastically thereafter. Assume a circular tunnel with support applying a uniform radial pressure p_i.

Initially everywhere

$$\sigma_r = \sigma_\theta = \sigma_z = \sigma_0 \text{ where } \sigma_0 \text{ represents the far field.} \tag{5B.1}$$

The elastic limit is represented by:

$$\sigma_\theta = A + B\sigma_r, \text{ where } A = 2c\cos\phi/(1 - \sin\phi), B = K_p \tag{5B.2}$$

In the elastic zone, assuming conditions of plane strain,

$$\epsilon_\theta = u/r = [\sigma_\theta - v(\sigma_r + \sigma_z) - \sigma_0(1 - 2v)]/E \tag{5B.3}$$

$$\epsilon_r = du/dr = [\sigma_r - v(\sigma_\theta + \sigma_z) - \sigma_0(1 - 2v)]/E \tag{5B.4}$$

$$\epsilon_y = 0 = [\sigma_z - v(\sigma_\theta + \sigma_r) - \sigma_0(1 - 2v)]/E \tag{5B.5}$$

So,

$$\sigma_y = v(\sigma_\theta + \sigma_r) + \sigma_0(1 - 2v) \tag{5B.6}$$

Hence, by substitution for σ_y in eqns (5B.3) and (5B.4), and by differentiating eqn (5B.3) with respect to r, we obtain:

$$\sigma_\theta - \sigma_r + (1 - v)rd\sigma_\theta/dr - vrd\sigma_r/dr = 0 \tag{5B.7}$$

and, from consideration of the stability of a segment (see Figure 5B.1),

$$d\sigma_r/dr + (\sigma_r - \sigma_\theta)/r = 0 \tag{5B.8}$$

hence, from eqns (5B.7) and (5B.8),

$$(1 - v)rd(\sigma_\theta + \sigma_r)/dr = 0, \text{ i.e. } \sigma_\theta + \sigma_r = 2\sigma_0 \tag{5B.9}$$

If $r = r_1$ at the plastic/elastic boundary, from eqns (5B.2) and (5B.7):

$$\sigma_{\theta 1} = A + B\sigma_{r1} = 2\sigma_0 - \sigma_{r1} \tag{5B.10}$$

Whence, from eqns (5B.9) and (5B.10),

$$\sigma_r = \sigma_0 - K/r^2, \sigma_\theta = \sigma_0 + K/r^2 \tag{5B.11}$$

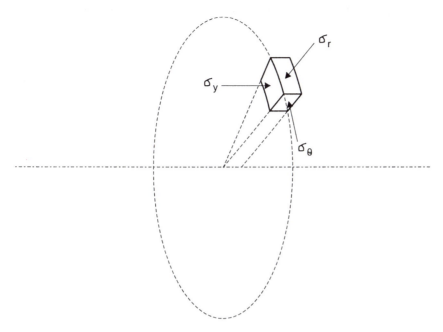

Figure 5B.1 Reference diagram for stresses around a cylindrical cavity.

and

$$K = r_1^2[(B - 1)\sigma_0 + A]/(1 + B) \qquad (5B.12)$$

Thus, from eqns (5B.3) and (5B.10):

$$u_1 = r_1(1 + v)(\sigma_0 - \sigma_{r1})/E \qquad (5B.13)$$

In plastic region, assume that:

$$\sigma_\theta = K_p'\sigma_r \qquad (5B.14)$$

In consequence, the equivalent to eqn (5B.8) now becomes:

$$d\sigma_r/dr + \sigma_r(1 - K_p')/r = 0, \text{ i.e. } \sigma_r = \lambda r^{(Kp'-1)} \qquad (5B.15)$$

where λ is a constant. Hence, at $r = r_0$

$$\lambda = p_0/r_0^{(Kp'-1)} \text{ and } \sigma_r = p_1(r/r_0)^{(Kp'-1)} \qquad (5B.16)$$

By equating σ_r at $r = r_1$ between eqns (5B.10) and (5B.16):

$$p_1(r_1/r_0)^{(Kp'-1)} = (2\sigma_0 - A)/(1 + B) \tag{5B.17}$$

Hence,

$$r_1 = r_0[(2\sigma_0 - A)/p_1(1 + B)]^{[1/(Kp'-1)]} \tag{5B.18}$$

If, following Kolymbas (1998), we assume volumetric dilation to comply with the yield function $\epsilon_v = b\epsilon_r$, since $\epsilon_v = \epsilon_r + \epsilon_\theta$, from eqns (5B.3) and (5B.4):

$$du/dr + u/r = b\,du/dr \tag{5B.19}$$

so

$$u = Cr^{[1/(b-1)]} \text{ where } C \text{ is a constant} \tag{5B.20}$$

but, where $r = r_1$, u_1 is given by eqn (5B.13) and r_1 is derived from eqn (5B.18).

Thus,

$$u = u_1(r/r_1)^{[1/(b-1)]} \tag{5B.21}$$

and, at $r = r_0$, $u = u_0'$ where

$$u_0' = u_1(r_0/r_1)^{[(1/(b-1)]} \tag{5B.22}$$

Closed-form solutions have been derived which are based on more specific properties of the rock but such problems are normally solved numerically.

If the ground is supported by a thin lining, thickness t, modulus E_l, in contact at all times with the ground, circumferential force per unit length of tunnel, N, is given by:

$$N = u_0 Et/r_0 = p_i r_0 \tag{5B.23}$$

or, for a continuous form of lining, E replaced by $E/(1 - v^2)$ in eqn (5B.23), providing a solution for p_i in terms of u_0 and hence, through eqns (5B.18) and (5.B22), a value for r_1. Figure 5B.2 indicates the stress pattern around the cavity.

If water is flowing through the ground of permeability k (see Appendix 5G) radially towards the tunnel at a rate Q, the loading on the tunnel lining is increased in two ways: (a) by the direct water pressure against the lining; (b) by stress transfer from water to rock as a result of radial reduction in water pressure.

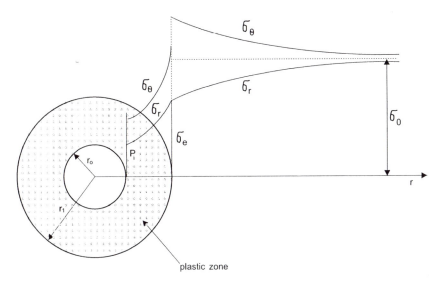

Figure 5B.2 Stress pattern for a cylindrical hole in elasto-plastic medium.

Assume ground of hydraulic permeability k, lining permeability k_l, mean lining radius r_0, lining thickness t, $t << r_0$, water pressure $w(w = \rho g H)$.

5B.1 Direct water pressure

In ground considered as a thick cylinder, external radius r_2, internal radius r_1,

$$dw/dr = -\rho g H Q/2\pi r k = -\lambda/r \text{ where } \lambda \text{ is a constant.} \qquad (5B.24)$$

So

$$w = -\lambda \ln r + A \qquad (5B.25)$$

where A is a constant.

If $w = w_2$ at $r = r_2$ and $w = w_1$ at $r = r_1$, from eqn (5B.25):

$$w_2 - w_1 = \lambda \ln (r_2/r_1) \qquad (5B.26)$$

for flow through the lining,

$$w_1 = \rho g Q t/2\pi r_0 k_l \qquad (5B.27)$$

assuming zero water pressure (w_0) within the tunnel. Otherwise all pressures need to be measured relative to w_0.

Hence, equating Q between eqns (5B.24), (5B.26) and (5B.27),

$$(w_2 - w_1)k/\ln (r_2/r_1) = w_1 r_0 k_l/t \tag{5B.28}$$

i.e. $w_1 = w_2/[r_0 k_l \ln (r_2/r_1)/kt + 1]$ (5B.29)

and Q may be evaluated through substitution for w_1 in eqn (5B.27).

5B.2 Water pressure transmitted through ground

For the purpose of illustration of the approach, we assume the ground to be elastic and to be unsupported at the ground/tunnel interface. This assumption greatly simplifies the working and is readily compensated as the calculations indicate. Equations (5B.3)–(5B.7) remain valid, with $\sigma_0 = 0$.

From consideration of the stability of a segment, eqn (5B.8) now has an additional term:

$$d\sigma_r/dr + dw/dr + (\sigma_r - \sigma_\theta)/r = 0 \tag{5B.30}$$

so, from eqn (5B.24):

$$d\sigma_r/dr - \lambda/r + (\sigma_r - \sigma_\theta)/r = 0 \tag{5B.31}$$

so eqn (5B.9) will now become:

$$(1 - v)rd(\sigma_\theta + \sigma_r)/dr = \lambda \tag{5B.32}$$

i.e. $\sigma_\theta + \sigma_r = [\lambda/(1 - v)] \ln r + A$ (5B.33)

At $r = r_2$, considering only the ground stresses caused by water flow, $\sigma_\theta = \sigma_r = 0$. If in addition, the ground has no radial support at $r = r_1$, eqn (5B.33) becomes, for $r = r_1$:

$$\sigma_{\theta 1} = [\lambda/(1 - v)] \ln (r_2/r_1) \tag{5B.34}$$

and, from eqns (5B.3), (5B.6), (5B.34),

$$u_1/r_1 = (1 - v^2)\sigma_{\theta 1}/E = \lambda[(1 + v)/E] \ln (r_2/r_1) = (w_2 - w_1)[(1 + v)/E] \tag{5B.35}$$

The radial stress between ground and lining, σ_{r1}, is now used in equating radial convergence for ground and lining at $r = r_1$ from eqns (5B.13), (5B.26) and (5B.34):

$$[(1 + v)/E][(w_2 - w_1) + (\sigma_0 - \sigma_{r1})] = (w_1 + \sigma_{r1})r_0/tE_l \qquad (5B.36)$$

where w_1 is solved from eqn (5B.29). It is of interest to note that if the lining were impermeable, the relationship would become:

$$[(1 + v)/E](\sigma_0 - \sigma_{r1}) = (w_2 + \sigma_{r1})r_0/tE_l \qquad (5B.37)$$

Appendix 5C Spherical cavity with internal support

The initial conditions are as Appendix B, also the yield criterion. Hence:

$$\epsilon_\theta = u/r = [\sigma_\theta - v(\sigma_r + \sigma_\theta) - \sigma_0(1 - 2v)]/E \qquad (5C.1)$$

$$\epsilon_r = du/dr = [\sigma_r - 2v\sigma_\theta - \sigma_0(1 - 2v)]/E \qquad (5C.2)$$

From considerations of stability of a spherical shell element (Figure 5C.1):

$$2(\sigma_\theta - \sigma_r) = rd\sigma_r/dr \qquad (5C.3)$$

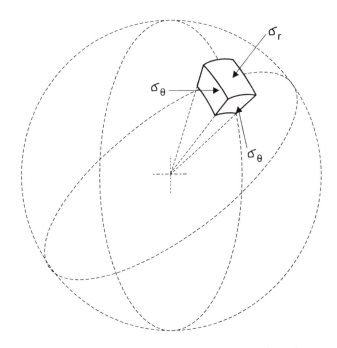

Figure 5C.1 Reference diagram for stresses around a spherical cavity.

Differentiating eqn (5C.1) with respect to r and subtracting from eqn (5C.2) gives:

$$\sigma_\theta(1 + v) - \sigma_r(1 + v) + r(1 - v)d\sigma_\theta/dr - rvd\sigma_r/dr = 0 \qquad (5C.4)$$

Substituting for $\sigma_\theta - \sigma_r$ from eqn (5C.3) gives:

$$r(d\sigma_r/dr + 2d\sigma_\theta/dr) = 0 \qquad (5C.5)$$

Differentiating eqn (5C.3) gives:

$$2d\sigma_\theta dr = rd^2\sigma_r/dr^2 + 3d\sigma_r/dr \qquad (5C.6)$$

From eqns (5C.5) and (5C.6):

$$rd^2\sigma_r/dr^2 + 4d\sigma_r/dr = 0 \qquad (5C.7)$$

Multiplying eqn (5C.7) by r^3 and integrating, noting eqn (5C.5), yields:

$$d\sigma_r/dr = K/r^4 = -2d\sigma_\theta/dr \qquad (5C.8)$$

i.e., $\sigma_r = -K/3r^3 + L$, $\sigma_\theta = K/6r^3 + M$ (K, L, M being constants)
$$(5C.9)$$

Where $r = \infty$, $\sigma_\theta = \sigma_r = \sigma_0$ so $L = M = \sigma_0$, hence eqns (5C.9) become:

$$\sigma_r = -K/3r^3 + \sigma_0, \ \sigma_\theta = K/6r^3 + \sigma_0 \qquad (5C.10)$$

and, from eqns (5B.2) and (5C.10),

$$K = [A + (B - 1)\sigma_0]6r_1^3/(1 + 2B) \qquad (5C.11)$$

and, by substitution for K in eqn (5C.9):

$$\sigma_r = \sigma_\infty - 2(r_1/r)^3[A + (B - 1)\sigma_0]/(1 + 2B) \qquad (5C.12)$$

In the plastic region,

$$\sigma_\theta = K_p'\sigma_r \qquad (5C.13)$$

and eqn (5C.3) now becomes:

$$2\sigma_r(K_p' - 1) = rd\sigma_r/dr \text{ or } \sigma_r = \lambda r^{[2(Kp'-1)]} \qquad (5C.14)$$

Where

$$r = r_0, \sigma_r = p_i, \text{ so } \lambda = p_i/r^{[2(Kp'-1)]} \qquad (5C.15)$$

From eqns (5C.12) and (5C.15):

$$p_i = (r_0/r_1)^{[2(Kp'-1)]} (3\sigma_0 - 2A)/(1 + 2B) \qquad (5C.16)$$

where A and B are defined in eqn (5B.2).

Appendix 5D The reinforced rock arch

A simple means of assessing the effectiveness of rock-bolting of highly joined rock is to consider the bolts as providing transverse prestress to the rock arch around the cavern or around the arch of a tunnel. If the principal stresses in the rock at failure are considered to be associated by:

$$\sigma_\theta = 2c \cos\phi/(1 - \sin\phi) + \sigma_r(1 + \sin\phi)/(1 - \sin\phi) \qquad (5D.1)$$

where $K_p = (1 + \sin\phi)/(1 - \sin\phi)$, the contribution of the stressed rock-bolts will be to increase σ_r at any radius r by $\Delta\sigma_r = p_i a/r$, where, by Figure 5D.1, $p_i = T/(m + n)$. The increased value of ΔJ_θ is:

$$\Delta\sigma_\theta = K_p Ta/(m + n)r \qquad (5D.2)$$

The total effect for bolts of effective length $(b - a)$ (Figure 5D.1) will be:

$$\int_a^b \Delta\sigma_\theta = K_p \frac{Ta}{(m + n)} \ln\left(\frac{b}{a}\right) \qquad (5D.3)$$

Where γ represents the unit weight of rock, the height H of the column of rock, outside radius a, supported by the rock arch will be given by:

$$\gamma H = [(b - a) + K_p Ta \ln (b/a)/(m + n)] \qquad (5D.4)$$

The principle relies on tight rock joints, possibly relying on the rock-bolts to establish this condition if rock adjacent to the tunnel is loosened during excavation.

Appendix 5E The brickwork or masonry arch

While few tunnels are currently designed in brickwork or masonry, many such tunnels survive from previous centuries. For the sake of simplicity, in this Appendix all such tunnels are referred to as 'masonry tunnels'. The engineer concerned with assessment or repair of masonry tunnels

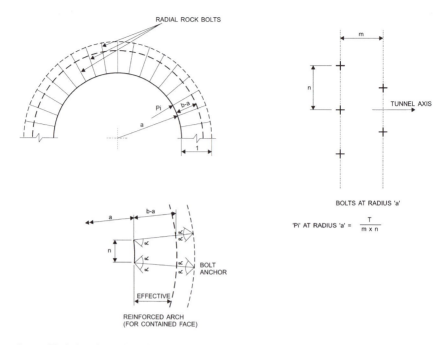

Figure 5D.1 Reinforced rock arch around tunnel (developed from Lang 1961).

needs to understand the appropriate criteria for establishing the stability of such tunnels.

The first requirement is that the structure should be in equilibrium with external forces, but this does not take us very far. Generally, a masonry arch may be considered as relatively flexible in relation to the ground it supports, on account of the mortar between the elements; loading on the arch needs to be compatible with the properties of the ground, taking account of strength and possible swelling potential.

Arches are normally statically indeterminate. Heyman (1982) recommends these basic procedures for assessment of the adequacy of the arch:

1. To check that sliding failure between blocks will not occur, i.e. that bedding planes, potentially weak in shear, are inclined at a high angle to the line of thrust around the arch (see, for example, Figure 5.3).
2. Tensile strength of the masonry should be neglected.
3. Compressive strength of the masonry is considered as infinite. While not a safe assumption in principle, provided the thrust line lies well within the width of the arch, the magnitude of thrust will not normally be a critical feature. For tunnels at considerable depth, this feature will need to be checked.

Heyman then restates the lower-bound theory of plasticity for a masonry arch thus:

'If a thrust-line can be found, for the complete arch, which is in equilibrium with the external loading (including self weight) and which lies everywhere within the masonry of the arch ring, then the arch is safe.'

The strength of the Lower Bound Theorem is that it is only necessary to demonstrate the possibility of accommodating such a thrust-line, not that this is necessarily the actual line of resultant thrust.

In a simple example, if an arch needs to satisfy a relationship, K_0, between horizontal and vertical ground stress, it will be necessary to establish that an elliptical thrust line whose vertical to horizontal axes are in the ratio K_0 may be inscribed within the arch. For the simplest instances, the demonstration of such a thrust-line may be undertaken manually. Computer programs are available for more complex or delicate situations (Harvey 1988).

Such an approach will demonstrate the nature of potential weaknesses to be explored for a masonry tunnel:

1. a horseshoe tunnel or other tunnel with a high radius to the sidewalls, particularly where any slight deformity may present difficulties in maintaining a thrust-line within the structure;
2. a horseshoe tunnel with invert arch will depend for stability on high bearing loads on the footings, to allow a continuous thrust-line to be inscribed.

Brunel's Thames Tunnel (Figure 5E.1) is a good example of a robust brickwork tunnel capable of accepting a wide variation in external loading. An ellipse transmitting all vertical thrust through the external walls, demonstrates a tolerance for $K_0 = 1.3$, whereas a pair of ellipses, with major axis vertical, each drawn to transmit loading through the central dividing wall tolerates $K_0 = 0.6$. Hence the tunnel would be stable for $1.3 > K_0 > 0.6$. The high quality of the brickwork and the mortar lend support to this view of considerable robustness.

Those unfamiliar with old underground structures should learn to treat the information on drawings, whose refinement in presentation may conceal optimistic expectations, with scepticism, the data only being accepted following physical confirmation. Computer analysis based upon assumptions of elasticity of the masonry (even when the program eliminates tensile stresses) should not be relied upon; these ignore the capability of creep of the structure to compensate for apparent areas of high stress.

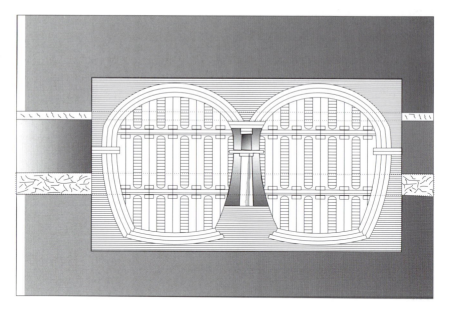

Figure 5E.1 Brunel's Thames Tunnel (after Law 1845).

Appendix 5F The ground model

A ground model, in the context of this Appendix, is understood to represent a statement of the characteristics of the ground on which design for stability of the tunnel is to be based. The model may be qualitative, e.g. a statement of geological structure, or quantitative, e.g. a time-based constitutive model of the behaviour of the ground – or any intermediate semi-quantitative model – depending upon the variability of the ground and the extent to which a more complex model may be justified by demonstrated benefit to the project. Muir Wood (1993) discusses the taxonomy of definition of rock support methods between these extremes.

As Duddeck (1981) rightly points out, the ground model is only one part of the overall design model, in which he recognises two phases (Table 5F.1).

It is the research model which explores the overall design problem objectively, selecting the dominant factors on a sound scientific basis, not by prejudgment. It is only by such a thorough approach that it is possible to distill the issues that may serve for an adequate 'technical model'. Without this rigour, it would be possible to match a theory to represent test results on a spurious basis. The process is further illustrated by Duddeck, indicating on a time basis the evolution of a design which represents the nature of expensive deviations which may be promoted either

Table 5F.1 Development of the design model (after Duddeck and Erdmann 1982)

1. Research model (portraying nature)	Explanation of phenomena Study actual loads and materials Analysis of parameters Establishing correspondence between theory and experiment
2. Technical model (replacing nature)	Developed for practical design Selection of dominant factors Idealisation of loading, physical characteristics and safety criteria No attempt precisely to model reality Lack of precise correspondence between theory and full-scale test accepted

by the inexperienced or over-enthusiastic theorist, or as an over-reaction to misinterpreted failure.

The first objective in selecting an appropriate ground model must be to identify possible failure modes. Failure need not be confined to structural failure, but may include unacceptable ground movements, problems in selecting suitable plant for construction, or excessive entry of water into the tunnel. These potential failure modes identify areas for study and are themselves modified as the studies develop, as part of the overall *design* process. As explained in Chapter 6, the nature of variability along the tunnel may determine the level at which to pitch the ground model. The fuzzier the ground model the less the benefit in attempting a good fit between the model and reality. The sharper the model, the greater the probability of an economic tunnelling system overall. The options may be expressed in order of increasing precision in the simplified terms of Table 5F.2. Each numbered factor is assumed to include all the preceding factors.

Table 5F.2 Types of ground model

1. Geological structure	Fundamentally a descriptive model which establishes limits of variability of salient factors
2. As (1) + simple qualitative factors	RQD or similar simplified representation of rock quality or selected relevant parameters for soil
3. As (2) + monitoring	Simplest basis for Informal Support
4. As (3) + quantitative design model	Adequate for analysis based on continuum–discontinuum or on elasto-plastic models of increasing complexity

It will be observed that the Rock Mass Classification Systems of Barton, Beniawski and others are omitted from Table 5F.2, as being inadequate to provide reliable information on the specific failure modes discussed above. See, for example, Moy (1995). All such systems are inappropriate for weak rock since none considers adequately rock strength, modulus or texture. Much effort has been applied to making statistical comparison between different Classification Systems, much less to establishing the reliability or economy of any one system except for a particular set of ground characteristics. The way ahead must be to identify the important parameters for a specific situation, and to present these in a multi-dimensional form as discussed in Chapter 2 and illustrated by Figure 2.7. The development of a data bank, which includes incipient and actual failures for different features of rock and rock structure will be a valuable aid to this approach.

Where a ground model of high precision appears to be justified by the delicacy of the operation, it becomes increasingly important to ensure that the manner of working and the standards of workmanship correspond to the assumptions of the model. Limitations in dependability of such aspects may lead to some modification of the ground model with the need for additional countermeasures, effecting a higher factor of safety, to compensate for suspected limitations.

Appendix 5G Ground-water flow into a tunnel

A granular medium is often assumed to have the same degree of permeability in all directions, i.e. to be hydraulically isotropic, unless it comprises layers of different textures or grain sizes, in which case the coefficient of hydraulic permeability will be orthotropic. This may readily be demonstrated, recognising that hydraulic permeability (simplified as 'permeability' below) is a term which takes account of the intrinsic properties of the medium and the properties of the fluid.

Suppose a horizontal layer of thickness p has a permeability k_1 and an adjacent layer of thickness $(1-p)$ a permeability k_2, for unit hydraulic gradient the horizontal flow will be $pk_1 + (1-p)k_2$ and hence the mean horizontal permeability will be k_h, where:

$$k_h = pk_1 + (1-p)k_2 \qquad (5G.1)$$

i.e. the weighted arithmetic mean of k_1 and k_2.

Vertically, however, for flow q per unit area, the head loss will be $pq/k_1 + (1-p)q/k_2$. Hence the vertical permeability, k_v, is given by:

$$q/k_v = pq/k_1 + (1-p)q/k_2 \qquad (5G.2)$$

i.e. the weighted geometric mean of k_1 and k_2. It can readily be shown that, where $k_1 \neq k_2$, $k_h > k_v$.

Traditionally, engineers are taught to estimate flow through a granular medium by means of a flow-net of potential lines and flow-lines. Flow-nets for an isotropic medium may be constructed approximately in 2-D, by sketches to maintain orthogonal relationships between the lines of the net. Computer programs are normally used for any but the simplest geometry, for anisotropic ground, and for 3-D flow, where axisymmetrical flow, where adequate, allows considerable economy in computation. The engineer should define precisely what is sought, including possible expedients for modifying the boundary conditions in order to establish appropriate forms of analysis.

The simplest approach to the calculation of flow into a circular (or an approximately circular) tunnel is to use the conformal transformation whereby ϕ represents lines of equipotential and ψ lines of flow, and:

$$\partial\phi/\partial x = \partial\psi/\partial y, \quad \partial\phi/\partial y = \partial\psi/\partial x \text{ where } z = x + iy, \ w = \phi + i\psi \quad (5G.3)$$

We can choose

$$\phi = \mu \ln (r_1/r_2), \quad \psi = \mu(\theta_1 - \theta_2) \quad (5G.4)$$

whereby r_1/r_2 = lines of constant potential represented by a series of circles, for flow between source and sink. If these are spaced $2h$ apart, H represents head difference between surface and tunnel, and Q_0 is flow per unit length, where r_0 is tunnel radius and h depth from rock surface to centre of tunnel, (Figure 5G.1) It is easily deduced that:

$$\phi \ (tunnel) = kH \quad (5G.5)$$

and total flux Q_0 is given by:

$$\phi \ (tunnel) = -(Q_0/2\pi) \ln (r_0/2h) \quad (5G.6)$$

where k represents hydraulic conductivity (permeability).

Hence;

$$Q_0 = -2\pi kH/(\ln 2h/r_0) \quad (5G.7)$$

If $h \gg r_0$, the equation is precisely correct. For low values of h/r_0, the correction factor is:

$$h(subst)/h = \sqrt{(1 + r_0^2/h^2)}, \text{ a correction of only 3\% where } h/r_0 = 4.$$

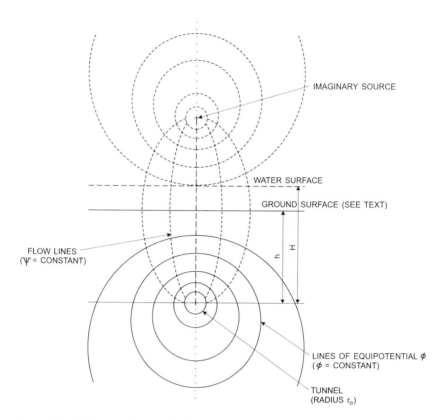

Figure 5G.1 Reference diagram for flow into a tunnel in permeable ground.

While engineers usually express hydraulic permeability (or conductivity) as a rate of flow through unit area under unit head gradient as a velocity measured in metres per second, another unit is the Lugeon, expressed as flow in litres per minute from a one metre length of a test hole between packers, under an excess pressure of 100 m head of water. The diameter of hole is not specified but normally considered to be NX size (diameter = 76.2 mm), which is found to correspond to about 10^{-7} m/s. There is often a need to relate the flow into a probe-hole for a tunnel to the expected flow into an unlined tunnel under corresponding conditions. From eqn (5G.7), for any thickness of permeable ground h, Q_0/Q_p (where the subscripts relate to tunnel and probe-hole respectively) is given by

$$Q_0/Q_p = \ln (2h/r_p)/ \ln (2h/r_0). \tag{5G.8}$$

For a value of h = 100 m, r_0 = 2 m, r_p = 20 mm, the ratio is found to be about 2. Of course, for high rates of flow into a probe-hole, the flow may be throttled by head losses within the hole.

Where k is calculated from pumping tests from a vertical well in a confined aquifer, the calculation of permeability is based on the loss of head H between observation wells at radii r_1 and r_2:

$$Q = 2\pi k H / \ln (r_2/r_1) \text{ c.f. eqn (5G.7)} \tag{5G.9}$$

Thus, pumping tests from wells or results from packer tests or other forms of tests in vertical boreholes will measure predominantly horizontal permeability, whereas a horizontal tunnel will depend predominantly on vertical permeability. For a packer test, the greater the test length, the less the extent of flow dependence on vertical permeability; hence for a truly homogeneous but orthotropic medium, the effect might be estimated by varying the test length for repeated tests. In order to apply results from test to prediction, it is necessary, therefore, to give some thought to the effects of anisotropy, the general condition for layered soils or jointed rocks.

Chapter 6

Design of construction

To the solid ground of nature trusts the mind that builds for aye.
Sonnets Part 1, No. 24, William Wordsworth.

6.1 The construction process

Tradition in tunnelling, as in other areas of construction, has entailed a differentiation between what is designated as 'engineering' and what is described as 'management of construction'. For successful projects for the future, in recognition of the increasing integration between functions, this barrier needs to be broken down. Of course, people will continue to be charged with specific functions to perform but the whole management structure needs to encourage appreciation of the interdependence of functions. Nowhere is this more necessary than for tunnelling involving application of Observational Design (Section 2.7) or 'Informal Support' (Section 5.2). The tradition of separation is so strong that inertia against change has caused design-and-build projects, potentially capable of full integration, to remain locked into outdated roles, and hence to perform sub-optimally. Overall success of the project is the shared goal, as against parochial success of one faction (possibly demonstrated by some performance indicator) at the expense of the subdivided responsibility of another.

During the construction phase of a tunnel, there are essentially three functions in progress affecting the physical execution of the work, as described and defined below. The unfamiliar terminology is deliberate, to avoid preconceptions as to the agency of each and thus to avoid assumptions concerning traditional roles of Owner (or Employer), Engineer and Contractor. The three functions are:

1. *Prediction.* Consideration of all the available relevant evidence should have provided the basis for the tunnel design (Chapter 5) and the designer as 'Predictor' will, in consequence, be predicting the behaviour of the ground and the tunnel in conformity with the assumptions upon which the design is based. It is important that the predictions

should include the full period of the construction process and any critical intermediate phases in the process. The predictor needs to specify clearly the features of workmanship essential for the success of the design assumptions made and any feature for special concern.

2. *Execution*. The construction needs to be planned to take account of the predictions, having regard to overall safety and security of the works, and economy of operation. Execution may well include: the undertaking of progressive exploration of the ground in phase with the advance of the tunnel; application of any special expedients required by the Predictor, in appropriate sequence with the progress of the works.

3. *Observation*. The term is used here in the most general sense, embracing inspection of the work carried out and the manner of its undertaking. Observation may also be required to provide design data for the future use of the Predictor, for example in permitting refinement of the initial design.

As implied by Figure 6.1 the procedure between Prediction, Execution and Observation is iterative with the constant attendance to the aims of controlling risk and of economy overall. Direction of the total process by the 'Conductor' (see Section 2.1.2) ensures balance between the contributions by the several 'players'.

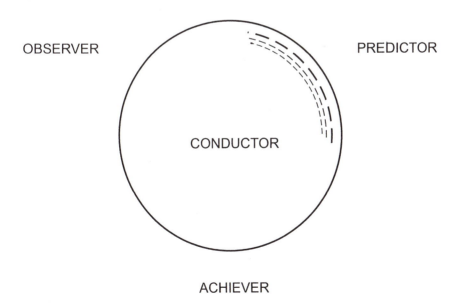

OBSERVER

PREDICTOR

CONDUCTOR

ACHIEVER

Figure 6.1 Functions of the tunnelling process considered as a system.

6.1.1 Prediction

Prediction will have more or less importance as a contribution depending on the nature of the tunnelling project. Thus, where the tunnel face is generally continuously concealed, for example by some form of pressure-balanced full-face machine, there will only be periodical changes in prediction, based on topography, punctuated by information from site investigation and inferences from the information provided by the Observer. At the other extreme, where the advance depends on interpretation of probe holes ahead of the face in relation to the need for special expedients, there may be frequent variation in the work plan, with constant need for reinterpretation of the predictions in terms of a practical plan for respecting the conditions for undertaking the work.

In such circumstances, economic construction will depend upon a plan of work prepared to different degrees of detail for different distances (and times) ahead, based on the progressive acquisition of knowledge, conceived in such phases as:

- *Phase 1*. Provision of general guidance for 3–6 months ahead to encourage planning for continuity and to ensure the availability of special plant and equipment ahead of requirement at the face. This period will also provide opportunity for planning of aspects of work which might otherwise entail potential conflict between different operations.
- *Phase 2*. Provision of guidance for a length (and period of several days) ahead for which information suffices to indicate specific departures from recent experience or of the need for special expedients. This will usually be achieved by exploration ahead of and around the face of the advancing tunnel by drilled probes, possibly supplemented by geophysical means.
- *Phase 3*. Provision of the most specific guidance for the immediate shift working. The exploration undertaken under Phase 2 supplemented by more detailed investigation of zones of special concern that may have been revealed, thereby establishing not only the delineation of such zones but a sufficiently precise nature of the problem to design the solution.

There is in consequence a phased qualitative improvement of predictive information which marches ahead (or should march ahead) in step with the advance of the tunnel. This process is illustrated by Figure 6.2 and needs to be pursued systematically, modified to suit changing circumstances but always retaining the objective of ensuring that work at the face can progress deliberately and methodically without delay (a feature that tunnelling shares with good military strategy and tactics), that might otherwise result from

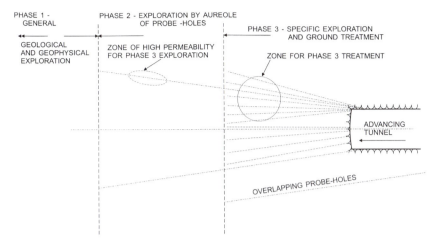

Figure 6.2 Phases in progressive exploration.

encountering an unexpected hazard. For example, where special plant or expedients may be required, advance planning should have taken account of the period needed for their commissioning, which may need to be reviewed as work proceeds in the light of new information causing perception of benefit from change to an initial scheme of working.

6.1.2 Execution

Tunnel construction methods may be classified in several ways. For the present and with the object of illustrating the differing needs for advance warning of potential problems, there may be distinctions in respect of the degree of 'robustness' across variations of the ground in relation to:

- *Tolerance*: the ability to operate without major problems within a wide range of variation of the ground;
- *Adaptability*: capable of modification, without appreciable cost or delay, to meet the foreseen variations of the ground.

As illustrated by Figure 6.3, there will be circumstances, represented by combined characteristics of the ground, favourable to efficient working and a wider range of such characteristics tolerable for a small percentage of the ground to be tunnelled. As an example, drill-and-blast will be a tolerant method for a rock tunnel, capable of accepting a wide degree of variation in rock characteristics and in geometry of the excavation, also in combination with a variety of special expedients. A shielded method of construction with TBM, capable of operating in closed (i.e. pressure balancing) mode or in open mode is an adaptable method, the nature of the particular degree of adaptation being related to expected variability of the ground. Figure 6.3 assumes that each method illustrated is limited by a value of ground strength q_u adequate for stability and by upper limits of q_u related to inherent strength and to RQD. There will of course be many other factors, with the potential of development along the lines of Figure 2.7 which relates to a different aspect of tunnel design. Tolerance implies an immediate acceptability of a certain range of conditions. Adaptability implies the ability to accept a certain range of conditions provided their anticipation permits appropriate adaptation in advance. Tolerance is thus a general attribute, adaptability one that is premeditated and requires a modification to the general scheme of operation. At the fringe, there is no sharp distinction between tolerance and adaptability. Thus, a drilling jumbo for a drill-and-blast tunnel will, with advantage, be itself adaptable for use in operations that are foreseen as being required for ground treatment or similar 'special expedients'. Exceeding the limits

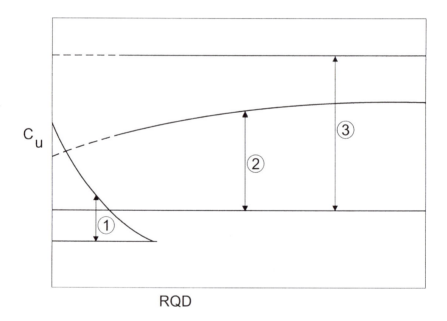

① HAND SHIELD OR SHIELDED TBM
② UNSHIELDED TBM
③ DRILL - AND - BLAST OR OTHER MEANS

Figure 6.3 Examples of tolerance of methods of tunnelling.

of acceptability for highly mechanised systems of tunnelling will entail high costs of dismantling and withdrawing machines coupled with subsequent loss of use and time.

Much information is now available on the practical performance of different types of TBM (Wagner and Schulter 1996). A general comprehensive report for the Deutscher Ausschuss für unterirdisches Bauen (DAUB 1997) provides a good starting point, indicating the several genera, species and sub-species, with the ranges of application expressed in general terms. Patrucco (1997) describes European Community development of technical standards for tunnelling machines. Bruland (1998) introduces Norwegian experience in modelling performance and cost of hard-rock machines. Biggart and Sternath (1996) draw on Storebaelt experience in indicating approximate ground conditions applicable to the use of slurry machines and EPB (Earth Pressure Balance) machines.

The practical limits for operating a slurry machine in soft ground may be related directly to the criteria for establishing face stability. Anagnostou and Kovári (1996) show how face stability is reduced by the extent of percolation of the slurry into the ground. If the permeability of a granular soil may be approximated as $k \propto d_{10}^2$ where d_{10} represents the grain size corresponding to the 10% smallest fraction of the soil, it is then possible to represent effectiveness of the slurry in terms of grain size. This allows, for any particular combination of soil type and ground water pressure, the construction of a diagram such as Figure 6.4 where z_0/a is plotted against d_{10}. For high values of z_0/a, a degree of overpressure at the head of the slurry machine may be tolerated and hence some increase in the acceptable value of d_{10} but efficiency reduces transverse to the line of the drawn curve. Anagnostou and Kovári also draw attention to the benefit for an EPB machine in increasing water pressure in the head of the machine above the pressure in the ground (or the alternative of ground-water lowering having account to possible risks of settlement) to increase the 'silo effect' in the support of the prism of ground ahead and above the face (Davis et al. 1980) and thus affecting its stability. All types of ground are variable to some degree and the choice between slurry or EPB machine will take account of reducing efficiency of the former with increasing values of d_{10} (with an absolute limit of $d_{10} \sim 2$ mm) and with reducing efficiency of the latter with reducing values of d_{10}.

For finer soils a machine operating on the principles of EPB will be preferred, apart from soils which are fine enough to 'stand up' and to have a low enough stability ratio, N_s (Section 5.1) to require no great degree of face support, for which an open face shielded TBM will then be preferred (Figure 6.4). Machines have been developed capable of conversion from one mode to the other and several machines have been used capable of operating in open or closed mode. For a specific project, the added efficiency offered by the ability to change modes needs to be set

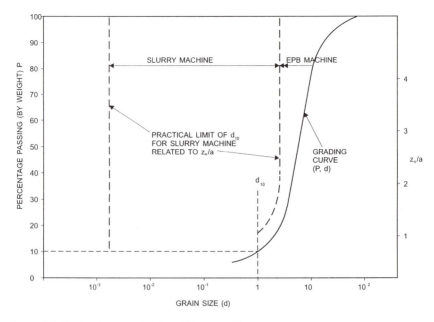

Figure 6.4 Choice between EPB or slurry machine.

against the added cost of a bi-modal (or tri-modal) machine, including the costs of delays in making each conversion during construction, under a range of circumstances based on geological assessment.

A great diversity of sizes, shapes and techniques of pipe- and tunnel-jacking is now available to the subterranean engineer (Thomson 1993). Comparable techniques for the emplacement of other structural features such as bridge abutments have also had wide application.

With the growing use of intermediate jacking stations (IJS), the length of a pipe-jack is only controlled by logistics and other factors of differential cost against alternative forms of tunnelling. The maximum drive length depends on the diameter of the pipe, say, 100 m for 300 mm diameter, 1000 m for 2000 mm diameter, with the possibility of including IJS where diameter > 750 mm. It is found to be a wise precaution to introduce a new IJS before the jacking force exceeds a certain proportion of the ultimate acceptable limit, to allow for unexpected increase in load. Curved drives of 100 m radius have been achieved with shaped ends to adjacent pipes and of 300 m radius using, for example, 1.2 m lengths of 1.5 m diameter pipe with timber packings capable of adequate distribution of load. The Unitunnel system inserts inflatable bladders at each joint, permitting by appropriate control to snake the tunnel, in Japan to 31 m radius for 1100 mm diameter pipes.

The different types of head adopt the features of many varieties of TBM, or of a shield with internal excavator for the larger bores. Where the diameter exceeds, say, 2 m, the operator may control the excavation from the head, otherwise this is undertaken remotely from a central control station. Compressed air may be used with pipe-jacking, observing appropriate controls and precautions. Since pipe-jacks often occur close to the ground surface, particularly for box-culverts, lubrication or the use of a membrane may be necessary to avoid local disturbance as a result of periodical adhesion of the ground to the upper surface of the box. Lubrication by bentonite may be an automatic part of the jacking process to reduce the load on the jacks and some form of external weak grouting may be necessary for pipe-jacks in rock to assist steering and control settlement. Pipe-jacking is often near the surface and may need to be able to cope with contaminated ground (Sharp and Turner 1989).

Many types of joint have been developed with the object of minimising damage by local overstress, maintaining alignment and preventing ingress of ground or water. Where the pipe provides a permanent conduit, watertight seals will be necessary at each joint.

A number of miniaturised features of pipe-jacking have been introduced as 'microtunnelling', first developed in Japan in the 1970s. By definition, the bore is less than 900 mm, the driving head remotely controlled, either using a pilot hole reamered out, an auger or by one or other form of pressure-balancing and excavation. When replacing existing pipes, a pipe-cruncher may be used, fragmenting the old pipe ahead of insertion of the replacement.

6.1.3 Observation

The process of 'observation' includes several functions which have traditionally become separated:

- *Inspection.* An operation which is intended to ascertain that the work is conducted in full compliance with the specification supplemented by other particular requirements, including those arising from 'prediction'. Aspects affecting risk that depend on temporary works and on workmanship should be specifically included in inspection, thus ensuring a safe scheme of working with, particularly for concealed work, a result of requisite quality overall.
- *Geological observation.* The methodical examination and recording of the ground (where such is practical and valuable) with special vigilance for features that may give advance warning of the unexpected, which may otherwise constitute an unforeseen hazard.
- *Performance observation.* Whether or not Observational Design (described in detail in Chapter 2) is adopted, markers for satisfactory

performance will include obvious characteristics of defects such as cracking or deformation and the methodical recording by instrumentation of movements, strains, possibly stresses and other features such as ground-water levels and pore-water pressures. Such observations are loosely referred to as 'monitoring'; correctly, monitoring implies that some remedial action may be taken if predetermined limits are exceeded. The advantage of an Observational Design approach is that work that may need to be undertaken as a result of monitoring has been designed and prepared in advance.

Inspection will generally be concerned with regular recording of compliance but it may have consequences requiring change, such as:

- concrete quality records, e.g. for strength, soundness and durability, which may indicate need for variation in quality controls in batching and production;
- procedures in construction which may need to be revised or specified in greater detail to avoid problems at intermediate stages.

Geological observation will make direct contribution to the specific information required for succeeding advances of the tunnel (as part of Phase 3 in Section 6.1.1), to any more radical change in approach, also in providing explanation of geological factors relevant to any local problem experienced (e.g. in distortion of tunnel support). Geological logs present a problem unless prepared in a highly systematised form, allowing, for example, a 3-D model to be prepared, and in providing readily accessible indication of special features of interest. Like a detective novel, geological logs may reveal clues which only later are seen to be advance indicators of the plot. For a large project, there may be hundreds or thousands of sheets of records which are of limited value unless capable of synthesis to indicate the wider picture and for extraction of specific features of engineering interest. For a creditably comprehensive account of geotechnical observations undertaken for a major project, the reader is directed to Harris *et al.* (1996), particularly to the chapter by Sharp *et al.* (1996).

As with all aspects of geological examination and reporting, as emphasised in Chapter 4, the geologist must be well briefed not only in the geological expectations – in order to react immediately to significant departure – but also to the geological features of greatest concern to the engineer and the uncertainties in such respects that remain from the pre-contract studies. This concern may arise as the result of association between two or more geological features, while one such feature on its own would cause no particular hazard. For example, unexpected tapering of an aquiclude, relied upon to prevent flow of water into a granular stratum, may give advance warning of previously unsuspected problems ahead.

Geological observations recorded for the project should always be related to features important, or potentially important, for the project. Where geological reports include special terms or geological jargon, care should be taken to ensure that the language and the engineering significance are explicit. The geologist provides expert 'eyes' for the project; time and effort should not be diverted to collecting data solely for the compilation of rock quality designation systems unless these are capable of being applied in a manner relevant to the project.

Performance observation introduces a wide variety of procedures and techniques, their design being an integral part of the overall construction process (Figure 6.5). At one extreme, performance observation represents little more than inspection by an engineer competent to notice subtle departures from expectation, e.g. in texture of concrete, in slight misalignment of segments, in incipient fracture in the roof of a rock tunnel, each of which may presage some feature requiring modification to the construction process. At another extreme, the performance observations may be the operative part of Observational Design whereby the need for premeditated supplementary work may be required. This procedure is described in Chapter 2.

Where observation may lead to action to guard against deterioration or damage, it is important to ensure that appropriate action is taken in a timely manner. Figure 2.9 represents that part of a typical organisation chart indicating the action needed to achieve this end. This is a feature which merits much care in accepting the management structure plan for a project. Even where there is nominally a single unit for procurement of design and construction in a design-and-build project, there may remain groupings related to identities of a company or company unit. These units may well cut across the information-and-action linkages needed to respond to a 'trigger' for appropriate response. A management chart may give indications such as 'interface at all levels' between the units. There must in addition be a specific linkage such that, whatever the occasion for responsive and possibly urgent action, the communication passes directly from the engineer who makes this assessment directly to the engineer

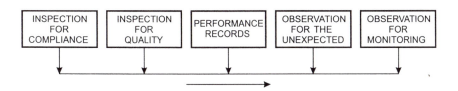

INCREASING DEGREE OF TECHNICAL DISCERNMENT

Figure 6.5 Observations of construction.

involved in putting this into effect (on previously rehearsed lines), with the event communicated to others who need to know, as opposed to a communication passing up the hierarchical line of the appellant and down the corresponding line of the activating agency. When considering the practicality of an observation-based approach, allowance for the time available for corrective action must include not only that required for taking the practical steps in the tunnel but also for the communication, reaction and issue of the appropriate instructions and the mobilisation of resources. The assumptions must be reflected in the organisational arrangements made in advance of such an event, combined with training to ensure that the procedure is understood and implemented without delay.

Observation during construction may lead to the conclusion, by those qualified so to rule, that the method of working may need modification to permit such techniques to be adopted. There are many examples of such circumstances, each pointing to the problems which arise when the contract conditions are inappropriate for the circumstances, particularly the degree of uncertainty of the ground. A few such examples merit recording here.

1. *Carsington.* The method of construction for this water supply tunnel in Derbyshire in limestone and shales of Carboniferous age had been specified in excessive detail with the Contractor effectively relieved of responsibility of certain of the most difficult tasks in treatment of the ground ahead of the face in the mistaken belief by the Engineer that this represented appropriate risk-sharing. In fact, the development of effective methods of drainage and treatment required free access to the face, prevented by the specified shields, road-headers and back-up equipment which needed therefore to be removed when serious problems develped. In view of the construction of the Contract, a major crisis resulted, leading to the replacement of the Engineer, in order to restore an effective approach to the work. The inappropriateness here resulted principally from inexperience in tunnelling of the consulting engineer appointed to this project, a failing which must be shared with the Employer in making this appointment without appreciation of the skills and experience required for the task. CIRIA Report 79 includes a specific 'health warning' to this effect:

   'There are few civil engineering projects in which greater ultimate responsibility weighs upon the Engineer than tunnelling. Those responsible for underground construction projects must be fully experienced in the practical aspects of recent tunnelling work.'

2. *Unidentified.* The Contract for an undersea outfall tunnel in strong jointed sandstone contained a provisional item for steel arch support.

The Contractor persuaded the Engineer that, to avoid delay, an advance order needed to be placed for this support. Having authorised this considerable expenditure, the Engineer subsequently agreed to these arches being built into the tunnel which, initially, required no such support. The tunnel was wet and the Contractor used the arches to support sheets of corrugated iron to deflect the water. Salt water caused rapid corrosion of the sheeting which formed an unsatisfactory umbrella, provided no support to the rock and obstructed inspection to identify potentially unstable rock blocks. Claims and counter-claims followed. Essentially there was an absence of understanding by the Engineer of the principles of tunnelling in such conditions and a determination by the Contractor to exploit this innocence.

3. *Unidentified.* A project for a coastal offshore outfall was expected to be situated in sedimentary rocks of Devonian age. Interpretation of a site investigation had concluded, in the absence of adequate consideration of geological structure and the significance of steeply dipping features between boreholes, combined with the positional relationship between boreholes and the tunnel (the alignment having been altered subsequent to the site investigation), that high strength sandstones would not be encountered by the tunnel. For the original intention of tunnelling by drill-and-blast, this assumption had little significance. The substitution to the use of a road-header should have led to a reassessment of risk of this occurrence, which caused problems of unacceptably low productivity and in consequence a very substantial claim which hinged on the question of foreseeability. A less tolerant scheme of tunnelling had been substituted for a highly tolerant scheme without appreciation that geological risk (R) depends on the product of geological hazard (H), the design of the project (D) and the design of the scheme of construction (C):

$$R = H \times D \times C$$

4. *Sewer tunnel in Cairo.* During the construction of sewer tunnels for the Greater Cairo Waste Water Project, one length of tunnel being advanced by slurry shield through water-bearing sand was found to be rising and distorting during construction. A first suggestion was that machine vibration might be causing local liquefaction of loosely packed sand. The tunnel was rising about 75 mm at the crown, twice as much at the invert, with the horizontal diameter increasing at axis level by about 150 mm; it therefore became clear that the tunnel was floating into the upper half of the cylinder of ground excavated by the TBM. Early in the drive, problems had been experienced as a result of the setting of grout penetrating from behind the shield into

the tail seal around the segmental lining. In consequence, a very weak bentonite-cement grout was being used in the vicinity of the TBM. Calculation readily disclosed that its shear strength was inadequate to resist flotation of the tunnel. The next problem was to determine why the sand in this locality was not, in such circumstances, settling onto the tunnel. After a suggestion that this might be explained by viscose from textile mills in the locality contaminating ground-water, a problem in shaft-sinking nearby provided the more likely explanation. Here tufa was found to have formed in the sand layer, doubtless caused by flow from the limestone Moharram Hills towards the Nile, providing cohesion between the sand grains to a degree adequate to allow stable arching over the tunnel. Use of a somewhat stiffer grout, subsequently confirmed by calculation as providing adequate shear strength, provided a simple solution to the problem. Might the site investigation have been expected to disclose this particular problem? The hazard only became a risk on account of specific and particular practices in construction.

6.2 The initial phases

6.2.1 Bidding strategy

Whether or not construction is separated contractually from design of the project, where competition between tenderers is primarily on price, a bidding strategy needs to be prepared. Assessment of risk must be a central component to such strategy, with geological risk the particular factor to stress for tunnelling. The assessment of geological risk at the time of tender may be approached in three elements:

(a) factual information of geological hazards available to tenderers which may be supplemented by information separately in the possession of a tenderer;
(b) interpretation of factual data with areas of major uncertainty identified in relation to engineering consequences, attached, where possible, to estimates of probability;
(c) consideration of the extent of the geological hazard imposed on the tenderer by the terms of the Contract, giving rise to geological risk when coupled with a preferred method of construction.

Bidding strategy will be dominated by factor (c) above. In this context, we may consider three possible degrees of risk sharing:

1. *risk sharing zero*: full imposition of geological risk imposed on the Contractor;

2. *protection against extreme risk*: a phrase in the Contract such as 'unless unforeseeable by an experienced contractor' or 'he could not reasonably have expected to be a significant probability';
3. *equitable risk sharing*: elements of geological hazard of major importance to a preferred scheme of construction and to its cost identified with reimbursement based on stated Reference Conditions (see Section 2.1.3).

For Case 1, in legal terms there is no limitation to the geological risk accepted by the tenderer, however dependent on information provided on behalf of the Owner. In consequence, there is no procedure for modification to risk in relation to the hazard assessment derived from factors (a) and (b) above. In principle, it would be oppressive if the Contractor were to be expected to take responsibility for circumstances incompatible with data provided, or with the interpretational limits made, by those working on behalf of the Owner (frequently only made available post Tender when preparing for litigation).

For Case 2, however reasonable such a formula for 'unforeseeable risk' may sound, there may remain considerable problems in interpretation, in the absence of an obligation on the Engineer (or Owner) to reveal his understanding of foreseeability *at the time of Tender*. The Owner should have been made aware at this time of the nature of areas of unforeseeability in relation to significant geological hazards which might eventuate, for which no express form of reimbursement is included in the Contract (and it should be made evident to the Contractor that this has occurred). Where such foresight has not been shown, or even made impossible by the nature of the relationship between Owner and Engineer, there are numerous examples of a weak Engineer falling to the temptation of implying, when circumstances are more unfavourable than he had hoped for, that the Contractor has been unduly optimistic in his interpretation of the data. For uncertain features which may have major effects on the possible options for the scheme of work, the satisfactory procedure is to provide a sum, in a manner transparent to Tenderers, for costs in dealing with the feature to the extent that it occurs. This has a considerable bearing upon the extent to which the Owner and Engineer need to study the implications of the working options prior to inviting tenders, an issue further developed in Section 7.1.

Where a contractor is concerned with bidding for a succession of several comparable projects in familiar circumstances, it is reasonable, extrapolating from experience, to select a provision for risk lower than the worst case for an individual project. Tunnelling projects are only rarely of this nature, however, each having its own degree of uncertainty and the individual project often of a magnitude to make the underestimate of risk burdensome.

There will be many features that will make evident to the tendering contractors whether the Owner (including the Engineer if a separate entity) has a professional attitude to the project and expects – and appropriately rewards – a correspondingly professional attitude from the Contractor. The Contract will then be based on Condition 3 with an explicit effort to make available all data that may affect the undertaking, with keys to the most vital data, also to have taken action to minimise uncertainties external to the project. The Contractor, in return, must be prepared to provide from time to time details of his proposals for undertaking the work in sufficient detail to permit assessment against any particular interests of the Owner which might be affected. There will then be a shared understanding of the nature of the problem and how best it may be circumvented.

A serious problem in tendering arises, as is illustrated by examples below, where a particular hazard may cause a scheme of construction otherwise preferred to become uneconomic. The dilemma is then whether to rely on this method and accept the consequent risk or, from the outset, to select a more 'tolerant' scheme. Under Condition 3 above, the Reference Conditions will serve as a basis for the decision; the experienced Engineer, who has drafted the Reference Conditions, will have assessed the overall risk and will have judged that it is sufficiently small to justify acceptance by the Owner (with the latter's knowledge and understanding of probable benefit and possible cost) of the potentially favourable but less tolerant scheme. Under Condition 2, the tenderer may indicate the nature of the assessed risk, provide a conforming tender based on the 'tolerant' scheme and an alternative (where this is permitted) on the basis that this specific risk is lifted from the Contractor. Under Condition 1, the tenderer is in a dilemma. Unless he wishes to depend on the result of fighting the case in Arbitration or in the Courts, he should provide in his tender for his most pessimistic assessment, possibly indicating a formula for an alternative scheme with a 'value engineering' cost saving by transferring part of the risk to the Owner, recognising that this will have minimal prospect of success if tenders are to be considered by those without a clear understanding of the uncertainties of tunnelling (an almost certain inference if Condition 1 has been adopted). In principle, a point emphasised in Chapters 5 and 7, Condition 1 is quite unsuited for tunnelling except in wholly familiar and predictable circumstances.

Where a tenderer accepts responsibility for design and construction, the problem becomes more complicated since the project design is a factor in relating risk to geological hazard; there should be opportunity, in consequence, for geological exploration specifically to explore important geological hazards prior to commitment to the preferred scheme of project design. While the general formula for bidding strategy remains much as described above, there will be additional factors in the risk calculations and additional elements of associated cost. These circumstances introduce additional problems in risk sharing unless there is effective sharing of complementary skills, perhaps integration, such as 'partnering', between the several Parties to the Project (Chapter 7).

During negotiations on qualified tenders there are often strong pressures for the tenderer to relinquish important qualifications. It is instructive for those concerned in negotiation to have access to a set of 'what-if' estimates indicating the approximate range of exposure to additional cost by acceptance of a particular risk. This discipline can dampen unwarranted enthusiasm on the brink of accepting intolerable risk as the price of winning a tender.

Special expedients (Section 6.4) may play a vital part in the scheme of construction in cost, in time – in the manner in which the tunnelling cycle may be affected, and in assumptions made about their efficacy. Performance claims by specialists need to be tested against actual achievement in comparable circumstances, with allowances made for differences. False optimism may be costly where, for instance, an elaborate scheme of ground treatment needs to be grafted into a system of tunnelling which has not made provision for such an eventuality from the outset.

The cynic can point to tunnelling contracts which have been set up on so unsatisfactory a basis, including in some cases the excessive loading of responsibility on the Contractor, that major renegotiation after a crisis has brought much consequent relief to the sitting Contractor. Reliance should not be placed on such a form of rescue.

6.2.2 The early phase of construction

It is a fact, an evident but inadequately appreciated fact, that many of the greatest problems, including the germinating seeds of future problems, occur in the early stages of the process of construction. These problems occur before familiarity with the method has been established by a newly formed team, before the effects of learning and training of any tunnelling project have been adequately absorbed and where no rhythm has been established, and yet at a time when there are great, often excessive, pressures for progress, especially from financial interests, with an eye to 'milestones' as a basis for payment, without understanding of the merits of a more measured and deliberate plan of operation.

A tunnel may be approached from a shaft, entailing an initial site set-up which will need to be modified before tunnelling begins. Whether the shaft is drilled, raised or sunk in a more traditional manner, the initial collaring and first stage of excavation need to be undertaken in such a manner to avoid unacceptable settlement, uncontrolled abstraction of water and any of the problems that may be associated with ground contamination.

One of the most potentially hazardous episodes in tunnelling in weak ground is that of the initial break-out from the access shaft. Whether or not the tunnel is to be constructed in shield or TBM, there will normally be (except in instances in which some form of gland or seal is provided, attached to the shaft wall to fit around the tunnel prior to the break-out) a period of potential exposure to disturbed, possibly water-bearing, ground which may flow at the face or around the perimeter of the shield. Where reliance is placed on ground treatment to control this risk, by grouts, resins or freezing, careful tests of its efficacy should be undertaken. Loss of ground into the shaft during sinking operations may cause voids against the shaft wall with possible effects on the stability of the shaft or the ground near the shaft/tunnel junction. Moreover, destruction of the original soil structure may result in a material difficult to treat (Section 8.2).

Unless a back-shunt has been constructed (and then the main problem is transferred to the break-out for this tunnel), a shield or other plant at the face will be deprived of its back-up by lack of space. Thus, for example, a shielded pressure-balance type of machine may need to be advanced a considerable distance before it is capable of being operated in this mode. Again, it may be stressed that this critical phase occurs during initial learning.

For the French section of the Channel Tunnel (Barthes *et al.* 1994), where all tunnelling was operated from a massive shaft, 55 m diameter × 65 m deep, the problem of protecting back-shunts and the initial lengths of tunnel drives were solved by constructing an approximately elliptical bentonite/cement cut-off, 198 m × 97 m in plan, toed into the Chalk Marl to envelope these elements of the early works.

Where a tunnel is approached from the surface directly to an entrance portal, the initial tunnelling will often be through ground decomposed by weathering, possibly water-bearing and affected by superficial movement, introducing problems atypical of the general features of the tunnel. Many stability problems have been associated with this initial phase, now frequently overcome by constructing some form of hood over the length affected, by means, for example, of an umbrella of drilled, driven or jet-grouted crown bars, the procedures described in greater detail in Chapter 5. A tunnelling shield or TBM, to be operated effectively, needs to be well buried in the ground, although forms of blade-shield may be more suitable for this initial phase. A more traditional expedient has been to construct the portal in a pit excavated from the surface, possibly over-size to allow subsequent traverse of a TBM or shield. Such a means is often adopted for road tunnels where, for visual or pollution reasons, the finished portal structure may then be extended well beyond the original surface of the ground.

The disposal of spoil from the tunnel may be a matter of considerable importance, meriting exploration by the Owner prior to inviting tenders. For hydro-electric projects, for example, the spoil may, with prior planning, be used in the construction of an associated dam – elsewhere it may be used for roads or for concrete aggregate. Conversely, the disposal of the spoil may present a major problem in mountainous country or where there are strict environmental restrictions on the location and topography of spoil heaps.

The Channel Tunnel provides examples in positive and negative directions. On the French side, there was no evident use for the Chalk Marl spoil. In consequence, at the main working shaft the spoil was slurried and pumped to disposal in a slurry pond behind a purpose-built dam on a hillside at Fond Pignon (Barthes *et al.* 1994). On the British side, inadequate space for the working site at the disused Shakespeare Colliery at the head of the main inclined access shafts was remedied as the work proceeded by using the spoil for reclamation as a series of cells, adjacent to the original seawall, bounded by a sheet-piled concrete seawall and dividing cell walls (Pollard *et al.* 1992). The phasing of the work required that in the early phases of tunnelling much improvisation in working was necessary until advance of the tunnels permitted an adequate area of reclamation.

London clay spoil from the Potters Bar Tunnels (Terris and Morgan 1961) was deposited, from skips using 'jubilee' track with limited compaction, on a large field, sloping towards the railway. Surface cracks indicated creeping downslope movement. Sampling allowed estimates of strength of the remoulded clay which indicated that the incipient failure would accelerate. Time nevertheless permitted construction of a brick rubble 'dam' ahead of the advancing toe of the spoil and hasty renegotiation of revised final contours for the field. The spoil came to rest against the 'dam'. Although born of a reaction to a surprise, the result was probably a more economic solution than the alternative of a high general compaction of all the placed spoil.

6.3 Choice of method

The most fundamental and irreversible (or reversible at great cost) feature of tunnel construction concerns the choice of method, from which flows all aspects of planning of the operation, of plant and logistics, and to which the whole concept of the project is closely linked. The choice of method may be dictated by the degree of certainty to which potential geological problems may be identified and located. This feature emphasises the need for planning of the studies which precede construction constantly to keep the construction processes in mind, as stressed throughout Chapter 4. Furthermore this interdependence adds weight to the deliberate step-by-step approach through the planning and design of the project (Chapters 2, 3, 4 and 5) to guide the studies. Benefits derived from innovation in method or from the need to reduce uncertainty in an unfamiliar locality may well justify the use of test galleries or experimental tunnels prior to decisions on the scheme of working, for which an allowance of time must suffice to permit the results to contribute to the optimal solution, recognising the period needed for observation and for possible modification of tunnelling techniques, as for the Keilder Tunnel (Ward 1978 and Coats *et al.* 1982), the Orange-Fish Tunnel (Kidd 1976) and the Victoria Line tunnels (Dunton *et al.* 1965). A more general account of the benefits is given by Lane (1975).

There are obvious, and not so obvious, factors associating the specific project with the means for its construction. The obvious factors include:

1. the nature of the ground, including scope for surprises;
2. lengths of continuous tunnel of a particular size, taking account of any requirement for relative timing of sections of the work;
3. extent of interconnection between tunnels;
4. useful tunnelled space in relation to practicable tunnel profiles;
5. value of time.

Other project-specific factors which may be significant are:

6. requisite spacing between tunnels;
7. local experience and maintenance facilities;
8. accessibility of project;
9. likelihood of late variations in requirements;
10. environmental concerns.

For overall project economy, there needs to be a clear mutual understanding, arising from discussion, of the significance of factors such as 1–10 above. Only then can the optimal construction strategy be devised. The possible significance of each of these factors is described below.

1. *Nature of the ground.* An elementary subdivision of the types of ground in relation to the scheme of tunnelling is described in Section 5.1, where the most important characteristics are expressed as a series of factors, such as intact strength and extent of jointing. These factors are incorporated selectively in the several schemes of Rock Mass Classification (RMC, Chapter 4). Different factors assume different degrees of importance in relation to different aspects of construction. For example, the common schemes of rock mass classification are relevant only to the needs for ground support, for which for example the Q system provides a rough initial notion of the support needs for strong, fairly tightly jointed rock, where say the Competence Factor R_c, [unconfined compressive strength of the rock]/[weight of overburden] > 4. Evidently the rate of deterioration of exposed ground, a

factor excluded from RMC, is important where rock bolts are expected to support the ground for an extended period. For shield and TBM tunnels, there may be problems of uneven loading on a shield prior to transfer of load to a solidly grouted formal lining.

Another factor of great importance concerns the variability of the ground, which may need to be considered more coherently than allowed by RMC. Thus, for example, when, as commonly occurs in weak sedimentary rocks, strata represent depositional layers or cyclothems, the coarser grained rocks (possibly sandstones) may be stronger and more permeable to water than the interleaved fine-grained layers (mudrocks or siltstones). These combined characteristics will tend to lead to problems of roof support and breakdown of the structure of the mudrocks of a nature and seriousness which might not have occurred with all of one type of rock or the other, unless expressly considered by the designer in shaping the tunnel profile (Sharp *et al.* 1984, Wallis 1998b).

The practicability and efficiency of all tunnelling processes depend on the reliability of assumptions made about the ground characteristics, which are adopted to form the several 'ground models' (Appendix 5F). Any attempt to over-simplify these characteristics in numerical form of RMC may serve to conceal the specific features which, in combination, present the critical features for design of the tunnelling process. The 'ground model' needs to express the known and the conjectured features of the ground, distinguishing between the two types, and how these features are inter-related. Variability must not at this stage be obscured by presenting mean properties of the ground. It is of the greatest importance that the design of construction should be based on the widest available knowledge of the ground. Elsewhere (Chapter 8), the practice, inspired by lawyers, of concealing from a Contractor factual reports on the ground and interpretative reports of site investigations, has been roundly criticised. Availability of information should be regarded quite separately from questions of responsibility for the information. There have been instances of assumptions made in error in early phases of interpretation which have coloured, without being questioned, the views of those who have followed and affected information passed to a contractor. In the absence of access to the development of such assumptions, a serious contractual dispute may arise.

Where geological features of special significance for tunnelling are likely to be preferentially orientated (see Chapter 4 for illustrations) these need to be taken into account. There are here two types of issue:

(a) Jointing may well be illustrated on a stereoplot (Figure 4.2) and hence indicate likely conjunctions of planes of weakness in relation to the line of a tunnel. If such data have been obtained from a series of vertical boreholes, allowance should be made for directional probability of intersection of an individual joint (which may include lost core being related to the presence of sub-vertical joints).

(b) When the feature in question is at intervals appreciably greater than the diameter of the borehole, the chance of intersection becomes small and hence may represent a hidden risk (see also Chapter 4 concerning deep weathered joints).

Where sharp geological change occurs within a length of tunnel, optimisation of the scheme of construction needs to pay attention to the particular features of the transition. If this occurs on an interface at low angle to the tunnel line, there may be appreciable lengths of mixed face presenting problems specific to the adopted means of tunnelling, e.g. the combination of hard and soft rock in the face of a TBM, throwing the machine off line while thrust pads sink into yielding ground.

2. *Lengths of continuous tunnel.* The classical example is that of a series of running tunnels for a Metro which traverse stations. What is the best order of construction? It is desirable to build flexibility into the programme by allowing either operation, i.e. of specific part of station and tunnel, to precede the other. There is often the need to assess the benefits of simplifying the tunnelling against the specific operational needs which will normally permit some tolerance in layout.

During the construction of the Channel Tunnel, the Crossover Tunnel under the sea from the British side in the Lower (Cenomanian) Chalk (Figure 6.6) was designed to be required to be advanced to a near complete stage before the TBMs for the running tunnels could be slid through for the completion of their drives, requiring the construction of a cavern of external dimensions 164 m long × 21 m wide × 15.4 m high. An alternative scheme, more flexible in relation to timing between the two aspects of construction, was considered and rejected as being more complex in organisation; it would have been less susceptible to the effects of geological uncertainty, a feature of some importance when excessive rate of local convergence was attributed to development of high water pressure some 4 m above the tunnel (John and Allen 1996), trapped above a relatively impermeable layer and causing the opening of a bedding plane. Figure 6.7 illustrates the French crossover tunnel in Chalk of approximately the same age, formed by an arch of multiple drifts around the previously constructed running tunnels (Leblais and Leblond 1996).

3. *Extent of interconnection.* Two issues arise here. Firstly, how do the interconnections aid the scheme of construction? Secondly, since

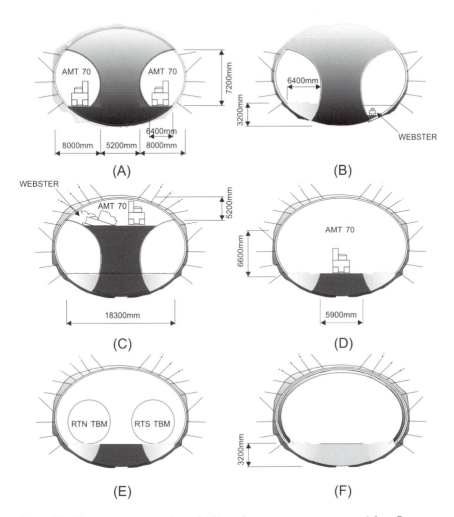

Figure 6.6 Phases in construction of UK undersea crossover-cavern (after Fugeman
et al. 1992).

the junctions between tunnels may provide particular difficulties in
construction, to what extent may their siting and their geometry
contribute to facilitating such problems. Section 6.1 discusses the principles of tolerance and adaptability in such respects. Interconnections
are often for reasons of safety in operation and hence their siting may
offer no great flexibility but, where possible, positions of particular
ground problems should be avoided. The Storebaelt Tunnels in
Denmark provide examples of the greater problems in constructing
interconnecting tunnels, deprived of protection by TBM, by contrast

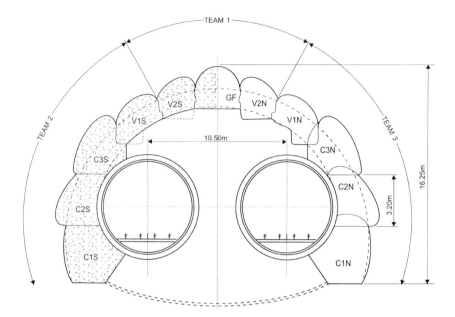

Figure 6.7 Phases in construction of French undersea crossover-cavern (after Barthes *et al.* 1994).

with the principal tunnels (Biggart and Sternath 1996). A tunnel designed to be watertight presents further problems to be solved at interconnections in safeguarding the integrity of an impermeable membrane at such points.

4. *Useful tunnelled space.* There is often the need for a certain profile for use, e.g. related to the structure gauge for a railway tunnel, with a need for an additional area of space of uncertain position or shape (within limits). Different means of tunnelling present differing, but always limited, options as to the internal profile of the structural element of the tunnel. This is another example of optimisation depending on the interplay between operational needs and constructional desiderata. A drill-and-blast tunnel or an Informal Support tunnel may readily make provision for local enlargements. For a TBM or shield-driven tunnel, the problems are greater, particularly where a segmental lining is employed in water-bearing ground. The preferable solution will be either to increase the diameter of the tunnel or to construct cells transverse to the tunnel. So far as the load-bearing shell of the tunnel is concerned, as discussed in Appendix 5E, it is important to ensure a continuous thrust-line around the shell and to remember that the compressive stress applied to the ground will be

approximately inversely proportional to the local radius of the 'thrust-path', hence highly concentrated at any sudden change in direction around the arch.

5. *Value of time.* For construction, there will be an optimal rate of working. Timing of a project should normally attempt to understand and respect this feature, making special provision where causes of uncertainty may justify a contingency allowance. There may be good reason for an accelerated rate of construction but this may require higher contingencies for the possibility of delays, since the exposure to the risk of delay will be increased and the designed slack in programme reduced.

6. *Spacing between tunnels.* Planning of a project may depend critically on an assumed spacing between parallel tunnels or tunnels crossing at low angle and minimal spacing. The spacing assumed prior to detailed consideration of the scheme of construction, if not capable of adjustment, may eliminate what would otherwise be the most economic means of construction.

7. *Local experience.* Systems of tunnelling do not necessarily export readily from country to country, for many reasons of culture, familiarity, resources, traditional ways of working, provisions for maintenance, not forgetting problems of importing special equipment and skills which those concerned with a country's import policy, remote from the project, may consider could be provided from within the country. Potential problems of importing the necessary resources of all natures should be carefully researched by those responsible for commissioning the project. Adequate provision should be made for training in techniques previously unfamiliar to the national or local work force.

8. *Accessibility of project.* Particularly where there may be dependence on special expedients, requiring rapid mobilisation or the risk of compensation of long periods of waiting time, accessibility and interaction between different operations may be an important issue affecting practicability and economics.

9. *Possibility of variation.* The different means of tunnelling have totally different characteristics in relation to response to change. Possibilities for change need therefore to be considered carefully, including those which may be prompted by a Party external to the project, e.g. an Authority concerned with aspects of health or safety, to avoid late variations or reconstruction. If such variation is foreseen as possible, what is the magnitude of the risk and on whom will the liability fall?

10. *Environmental concerns.* A far-sighted Owner will have undertaken preliminary investigations with Authorities whose jurisdiction may impinge upon the project, for environmental or other reasons. The Owner will also have consulted the community to establish understanding for the nature of the proposed project. The objective must

be to reduce to a reasonable level, for agreement in detail as the project proceeds in relation to the specific proposals for construction, the uncertainty affecting conditions of noise and pollution to be respected.

Limits on vibration are based on the combination of particle velocity and frequency at the affected point, related to the risk of annoyance or causing structural damage (Broch and Nilsen 1990). Where caused by blasting, the maximum particle velocity is roughly proportional to $Q^{1/2}/R$ where Q is the weight of the charge (detonating simultaneously) and R the distance from the detonation. Site specific factors, including the nature of the ground will affect the transmission of noise and vibration, with small trial charges usually employed for calibration and to establish means for compliance with the national or other standards. Hillier and Bowers (1997) describe vibrations caused by mechanical tunnelling plant, establishing personal sensitivities to vibration (greater vertically than horizontally). For Metro tunnels, there may well be concern for limiting transmitted vibrations during operation, where the solution may include elastomeric track suspension (Jobling and Lyons 1976).

6.4 Special expedients

Traditional means for tunnelling through bad, i.e. weak and water-bearing, ground entailed the use of low pressure compressed air or of ground freezing. LP compressed air has now been largely replaced by other expedients to stabilise the ground, although up to a pressure of 1 bar (100 kPa or about 10 m of water) there should be no effects on the health of those passed medically fit to work in compressed air; locking in and out are then fairly rapid processes. Compressed air may not only help to stabilise ground exposed at the face but also to reduce water inflow and, by reducing pressure differential between the ground *in situ* and the interior of the tunnel, the value of N_s (Section 5.3), to reduce ground settlement. While compressed air has usually been used for segmentally lined tunnels, on occasion it may with advantage be used with an 'informal' form of initial tunnel support, e.g. with rock bolts and sprayed concrete, the compressed air contributing to the natural arching in the ground around the tunnel (Appendix 5D).

The installation of air-locks may be possible within a tunnel, prior to the face reaching the ground for which the air is required, in which case, except for large tunnels, an enlargement will be required for the air-locks and adjacent sidings or laybys. More often the tunnel air-locks have themselves to be built under compressed air, entailing the use of vertical shaft air-locks with the consequential restriction on throughput of material, plant and people. An obvious consideration of the siting of air-locks is that the entire length of tunnel to be subjected to internal pressure of compressed air must have adequate ground cover or surcharge to ensure stability. For this condition, $N_s < 1$ (Jones 1998).

Marchini (1990) provides a wide-ranging introduction to the special expedients available to the tunnel engineer (See also Cambefort 1977). Special expedients of ground treatment by injection may be difficult to undertake for the initial length of a tunnel since there needs to be sufficient protection by rock (or possibly concrete) wall of the excavated zone in order to collar the drill-holes, taking account of the pressures to be developed during the treatment processes and the extent of the ground to be subjected to raised pressures.

The objectives for any use of special expedients need first to be thoroughly understood. Is the objective to improve the general strength of the ground, or of discontinuities in the ground by a process of consolidation, or is it primarily to exclude water? If the object is to exclude water, for what reasons and to what standards of residual inflow under what conditions of exposure of the rock? Often the concern is with water causing weakening and possibly flow of highly fractured or weathered rock or of fine soils into the tunnel. Where the water is travelling through a contiguous aquifer, the most effective means may be by combining the interception and drainage of water entering the aquifer with grouting, rather than by any attempt to tackle more directly the finer soils forming a potentially unstable aquiclude which will be more difficult either to drain or to grout.

During construction of the Clyde Tunnel, a sheet-piled box for the North Portal and Ventilation Building was designed to be sealed into glacial drift (boulder clay). An observation of air bubbles rising through the water in the partially drowned box implied that the seal had not been fully achieved. Investigations revealed that, while initial boreholes had proved the clay at two diagonal corners, a previously unsuspected glacial stream had cut through the clay on the other diagonal. After unsuccessful attempts directly to drain the silt layer overlying the drift, which might otherwise have flowed into the box as excavation proceeded, it was found that pumping from the permeable sandstone underlying the drift was effective in generally lowering the water-table in the silts. The construction of the portal proceeded without further incident.

Another incident of the Clyde Tunnel involved the driving of the tunnel under compressed air through a gravel esker reaching close to the ground surface (Morgan et al. 1965). It was appreciated that the pressure of compressed air could be transmitted to the top of the esker and in consequence lift the ground and cause a 'blow'. The solution adopted was to form a box from the surface of grouted bentonite/cement walls as the sides and lid of the box in the gravel, and to use compressed air at a low pressure supplemented by pumping from tubewells in the invert of the pilot tunnel. An air release valve was also fitted to control maximum air pressure towards the top of the esker.

Inflow into a tunnel may be objectionable *per se* in its effect on plant, on working conditions generally, on problems of disposal of the water or on weakening of exposed strata. Inflow may be objectionable on account of external consequences, as described in Chapter 5, for example on affecting local water supply or on settlement of buildings or structures. It is important to be able to place concrete linings so that the setting concrete is protected from direct contact with inflowing water. For many years, techniques have been adopted using continuous plastic sheeting attached to the rock around the arch of the tunnel, with piped drains in shallow trenches along each side of the tunnel invert; on completion of lining, inflow is stopped by plugging and grouting the side drains (Figure 6.8).

Drainage of an advancing tunnel by means of a sub-drain was a familiar technique of the nineteenth century (Simms 1944) but the use of well-points, possibly with recharge wells to avoid consequential damage, is a more widely used expedient at the present day for lowering the water-table in the vicinity of a tunnel. The most remarkable recent example of this technique was 'Project Moses' for the Storebaelt Tunnel which lowered the water-table in the Glacial tills under the sea by pumping from fissured

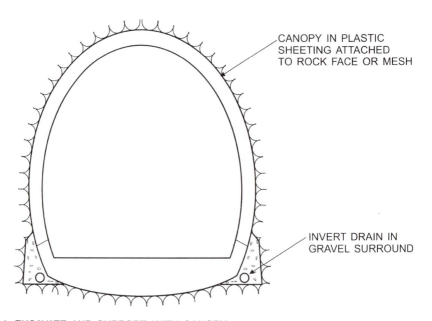

CANOPY IN PLASTIC
SHEETING ATTACHED
TO ROCK FACE OR MESH

INVERT DRAIN IN
GRAVEL SURROUND

1. EXCAVATE AND SUPPORT, WITH CANOPY
2. ESTABLISH INVERT DRAINS
3. PLACE CONCRETE LINING IN INVERT AND ARCH
4. GROUT UP INVERT DRAINS AND SURROUND

Figure 6.8 Concrete lining in wet rock.

limestone beneath (Biggart and Sternath 1996). The lack of adequate recharge wells resulted in considerable damage to elm piles (whose life depends on remaining saturated) supporting buildings in Amsterdam during the construction of the first line of the Metro.

A grouting scheme for granular soils or for rock needs to be designed for the specific circumstances and objectives, with a high degree of dependence on the size of aperture (for soils) or fissure (for rock) through which the grout is required to penetrate. The effectiveness of a suspension grout based on cement is related to the ratio of fissure width to size of cement particles. Thus, a grouting system based on Ordinary Portland Cement may be able to achieve an effective permeability of $k = 10^{-7}$ m/s but 5×10^{-6} m/s is more generally achievable (see Appendix 5G for further explanation). Grouts based on more finely ground cement, such as Microfine Cement, may perhaps achieve $k = 5 \times 10^{-8}$ m/s. For greater penetration one turns to chemical grouts (Harding 1946, Glossop 1968). Lichtsteiner (1997) surveys a range of cement and synthetic resin grouts. The traditional silicate grouts have typically low strength and relatively low life and hence have been widely used for shaft sinking through permeable sediments near to the ground surface. Acrymalide grouts are claimed to be able to achieve flow rates as low as equivalent to $k = 10^{-9}$ m/s but these grouts are toxic as the experience at Hallandsäs has emphasised. The Hallandsäs road tunnel in South Sweden, advanced by drill-and-blast, used acrylamide resin grout to prevent water inflow from depleting groundwater. Evidently, some form of syneresis occurred, allowing the toxic resin to reach the surface, resulting in the death of several cows (*Tunnels Tunnelling Int.*, November 1997 p. 8 and May 1998 pp. 22–4, Anon, 1998). Acrylic grouts are claimed to achieve similar results to acrylamide grouts but with somewhat lower toxicity.

Where grouting in rock is related to a specific localised feature of limited extent, the scheme is designed expressly for this purpose. More generally, a scheme of grouting for a tunnel in fissured rock will usually entail a succession of aureoles of grout-holes drilled at an acute angle to the tunnel in order to treat a series of overlapping cones of rock ahead of the advancing face. Where the degree of opening of water-bearing fissures varies, a thick grout may first be used to fill the wider fissures prior to using a more expensive viscous grout for the finer fissures. Coarse grouts may be used under high pressure to cause deliberate *claquage*, i.e. the temporary opening of fissures, but such an effect does not extend to a distance of many grout-hole diameters. Grouting is often undertaken through *tubes-à-manchette* to enable the grout to be delivered to a preferred zone. The processes of consolidation and compensation grouting are described in Section 5.3.

Ground freezing, usually to exclude water, has often been considered as an expedient of last resort on account of cost and time. Freezing, first used in shaft sinking (Chapter 1) was undertaken by the circulation of brine through U-shaped tubes in bore-holes, set in an annulus around the shaft. For tunnelling, liquid nitrogen may be used as the freezing medium, achieving a faster freeze and hence possible to use where there is some slight movement of ground-water (Chapeau and Dupuy 1976). The freeze pipes are coaxial, the liquid nitrogen delivered through the central tube with sufficient clearance at its lower end to avoid obstruction by the formation of ice. The external tube is partially insulated in order that freezing is confined to the zone to be treated. Freezing is generally undertaken from the surface but the process may also be operated from a tunnel, provided it is of a size to house the plant required for the operation. Then, the freeze pipes may be established by overlapping aureoles, as at the Milchbuck Tunnel in Zurich (Bebi and Mettier 1979; Jones 1996) or, to deal with a more confined zone of unstable ground, from the surface, as for the Three Valleys Tunnel (see also Section 8.2). Where freezing is used for a fine silt, ice lenses may form with deleterious consequences to the tunnel on thawing. Freezing will tend to break down the structure of clay, whose strength will be correspondingly reduced. The consequences of such phenomena need to be studied in relation to the stability of a tunnel through frozen ground (Altounyan and Farmer 1981). Considerable data have been published at international conferences on ground freezing, including the effects of freezing on geotechnical properties.

Chapter 7

Management

If a man will begin with certainties, he shall end in doubts; but if he will be content to begin with doubts, he shall end in certainties.
Advancement of Learning, Francis Bacon.

7.1 Introduction

It is remarkable that, with so great a recent emphasis on management, with much purveyance of courses, books and seminars, so little wisdom appears to emerge on the essence of the successful management of large complex engineering projects. Major tunnels clearly figure in this category, occupying a special place on account of the dominance of the construction options and the constant vigilance demanded in respect of geological uncertainty. The principal defect of so many tracts on management is that the subject is discussed as administration, the manipulation of the tools of management, understanding the bureaucratic machine, in preference to management as the art of blending and synthesis across the diverse contributions to the successful project. There is a great deal of jargon relating to project management. This language needs to be understood, in order to penetrate the surrounding mystique, but not to be used. Management-speak is no substitute for good leadership and clear thinking.

Management as administration supposes that the engineering is controlled by directives and undertaken in individual cells, each cell concerned with a particular aspect which is defined and recorded. Administration endeavours to police each aspect to prevent change which might otherwise interfere with other aspects of the project. Administration is remote, avoids technical debate, being incompetent, on account of inadequate technical understanding and an inappropriate structure, to engage in interactive leadership, reacting ineffectually to the consequences of change without active engagement in their anticipation.

Management as administration is practised by some of the best known management consultants, who rely upon a formal set of procedures to ensure rigorous compliance with each aspect of a project to avoid interference with any other aspect. Superficially, the consequence appears to be that of successful management, but achieved at great cost in denying the opportunity for change as advantageous options come to light as a project develops. In this way, the effective operation of the process of *design* is denied (see Chapter 2).

This widespread set of errors in project management springs essentially from a legalistic approach to project management which derives from the thesis that interests of participants are only guarded by precise definitions of each transaction or undertaking treated separately from any other. In essence, this approach endeavours to impose certainty in an inherently uncertain environment. In the terms of games theory, a zero-sum game is imposed, whereby the gains of one party can only occur at the expense of another. On the contrary, as described in Section 7.2, the essence of the successful project is the recognition that the parties must share to some degree the benefits of a successful project implicit in the process of optimisation, rather than reliance on scoring only by diminishing the benefits of other parties of the project, to the detriment of the project overall.

Successful project management hinges upon the application of the professional element of engineering. While engineering is a market-led occupation, the professional mark of an engineer's training should ensure that the interests of his (or her) Client, (i.e. the rewarder of his contribution) and of the public interest (the 'stakeholders', the social and environmental consequences) should predominate over the reliance on short-term market values of so many politicians and accountants. Such an attitude should also preclude totally the advancement of the interests of the engineer at the expense of his Client.

As the environment of underground construction has become increasingly exposed to commercial pressures so has it become increasingly difficult to retain professional standards – but not impossible. Excessive commercial pressures, failing to appreciate the merits of a *gestalt* or holistic approach (i.e. the perception of the project as a whole) are bound to achieve sub-optimal results. In the distant past, there was a presumption that the engineer concerned with the conceptual design of the project (in many countries, the Consulting Engineer) had virtually a monopoly of professional virtue. This never was the case and a present boost to professionalism results from the present sheer complexity of engineering, so that professional engineers are now fairly evenly spread across the more enlightened Owners, Engineers, Contractors and Specialists in plant and special processes. Many need, however, to relearn the demands upon a professional to perform effectively. The professional understands that his own expertise and know-how need, to be effective, to inter-relate with those corresponding features of others with whom he works. This, as indicated by Section 2.1, is an essential element of *design* and, in consequence, of the effective manager.

This chapter is concerned with establishing the criteria for good practice. There is much to learn from multitudinous examples of bad practice but these are best explored against their consequential effects and typical examples are therefore examined predominantly in Chapter 8. Not all – but most – project disasters derive essentially from management deficiencies, often of structural rather than personal nature. Too often, immediately evident and practical causes are discussed without appreciation, possibly even with deliberate obscuration, of the way in which the management philosophy or structure has encouraged or even established the deficiencies in practice, the project 'climate' (Pugsley 1966). Comparable to the way in which, as described by Pugsley, features of a project contribute to a 'climate' for physical failure, so also is there a less tangible 'social climate' of relationships, competencies and responsibilities that contribute to the broader features for success of a project, dependent, as is a tunnel, on a degree of inherent uncertainty. The wider concern is therefore with contract practices which give poor value for money, deny development of engineering responsibility and fail to exploit the potential capabilities for success.

7.2 Project procurement

The first step in the assembly of the component parts for the execution of a successful major project is to establish the means of procurement, i.e. the assembly of the contributory elements. It should be evident – but too often set to one side in favour of established procedures without merit for an underground project or for any project with a scope for change or uncertainty – that the purpose should be that of harnessing most effectively the special skills of those upon whom success must depend. A vital element is that such parties should be engaged in a professional manner, so that the success of the project dominates the shared purpose, with this success reflected in the contribution to the profitability of each participant.

For the Øresund Link between Copenhagen and Malmo, a deliberate attempt has been made (Reed 1999) by the Owner, Øresundskonsortiet (ØSK), to demonstrate the functions of an exemplary Employer (or 'Enlightened Purchaser' in the terminology of the UK Treasury), with procurement rightly seen as the first step, an essential basis for supporting the subsequent phases of a potentially successful project, of a size and complexity comparable to that of the Channel Tunnel.

Apart from the rigour with which the prequalification of tenderers and the award of Contracts was undertaken, to a preconceived plan for objective assessment of resources, competence and the degree of conformity to the standards of requirements, other particular features merit mention:

- While each of the four main contracts was conceived as design-and-build to give each Contractor greatest freedom in the method of working, an Illustrative Design was prepared as part of the Tender documents.
- Uncertainties in provisions to be made for extreme conditions of sea and weather (e.g. the extent of delay as a consequence of sea ice) and in the properties of the ground to be considered by tenderers were confined by means of 'Reference Conditions' (Section 2.1.3).
- An important feature of tender assessment concerned the proposals for fulfilling the Contractor's acceptance of responsibilities for the quality of the work.

The Illustrative Design not only provided the Owner with a basis for establishing overall feasibility, with preliminary estimates of time and cost, but it also permitted negotiation with the several Danish and Swedish Authorities responsible for planning, navigation and the environment prior to the work beginning, with much benefit in identifying and resolving issues that might otherwise have delayed or modified the work. Requirements were based largely on performance criteria, with the Illustrative Design provided for guidance and indication of standards of acceptability.

In order to establish overall credibility for technical project competence, the Owner absorbed elements from consultants, Rambøll, Scandiaconsult, Halcrow and TEC. In order to emphasise the intention to deal objectively with contractual issues that might arise during the course of the work, a Dispute Review Board of independent engineers was appointed for each main contract, kept informed on the progress of the work as the first external agency for appeal. Much thought was also given to the delicate problem of managing interfaces between Contracts, exploiting the opportunities for mutual benefit in cooperation, eased by bonus payable in achieving the overall project target and objectives. The conclusions of Reed in relation to this project merit repetition as indicators of the functions of a far-sighted Owner:

- ØSK acted as a very efficient buffer between the politicians and the engineers, and between the environmentalists and the engineers. This enabled the engineers to concentrate on what they are good at.
- ØSK had clear objectives for what they wanted to achieve, and how it should be achieved. This made it simpler to prepare the contract documents.
- ØSK recognised the strengths of the consultants and contractors who worked with them, and tried to ensure that those who were best able to handle a particular aspect did so.
- ØSK were always prepared to listen and consider suggestions put to them. There were no closed minds.

- ØSK realised that if one of their contractors had a problem, ØSK had a problem. This attitude had a significant effect on the content and the tone of the documents.

There are too many recent examples of faulty systems of project procurement to merit invidious selection. Identifying characteristics of the doomed project – doomed that is to suffer unplanned over-runs in time and cost usually accompanied by satisfaction for the legal vultures – may be summarised as entailing several of these features:

- risk imposed on Contractor in preference to adequate risk analysis and control;
- uncertainty of composition of project and performance criteria;
- absence of assurance of feasibility;
- absence of prior agreement with regulatory and other authorities;
- site investigation inadequately related to construction and not treated as a project resource available to participants;
- evidence of intention to rely upon commercial relationships enforced by law in preference to professional relationships built on mutual trust.

It is also to be noted that international financing agencies and Development Banks are not free from some of the above defects in the conditions that they impose.

The Øresund Link project provides an example for the future for a contractual base which appears to achieve the objectives of 'partnering' without losing the competitive edge and without dependence on a cumbersome preliminary procedure. It is not alone but too many of the other recent examples of enlightened management have arisen as a result of a major incident emphasising the deficiencies of those contractual conditions initially adopted.

7.3 The 'zero-sum' fallacy

A zero-sum game is one in which gains are only made by one party at the expense of losses by other(s). For example, if one team wins the Calcutta Cup, the other loses. The essence of game theory is to establish the optimal strategies between perceptive participants that will allow outcomes favourable to each.

This notion derives from game theory which was the creation of John von Neumann (1903–57), the renowned Hungarian mathematician who migrated to the United States in 1930, the theory first appearing in print in 1944 (Neumann and Morgenstern 1944). The significance of game theory is that the best ploy for an individual in any transaction depends on the actions of others, i.e. that the action of one depends on the perception of others as to their best interest. The application of game theory has exposed that, for a wide range of applications [e.g. in the biological domain, see Ridley (1996)], the optimal results are obtained through a degree of cooperation, as opposed to each attempting to take the path of maximum apparent self-interest.

In the context of tunnelling contracts, game theory provides theoretical support to the practical benefits of flexibility and cooperation as opposed to domination and rigidity imposed by the author of the contract. The nature of uncertainty is a key element, with benefit of provisions for resolution made to the advantage of all concerned. The overall gain over a zero-sum approach (e.g. the Channel Tunnel) may be highly significant.

The most dominant philosophical basis of bad management is the notion of the zero-sum game, a defect which springs from the dominance of lawyers, and those who have become dominated, by a process of diffusion, by the legalistic outlook on construction procedures. The only direct experience of most lawyers with construction is confined to negotiation of liabilities for restitution when something has gone wrong and blame is to be attributed retrospectively. A loss has been incurred and somebody needs to be held liable and to compensate another party. This is a situation in which the zero-sum concept is applicable, where gains equate to losses (provided for the present that the legal costs are excluded); it is then too late to mitigate the cause of the problem. It is from this process of sweeping up the pieces that the lawyers acquire the zero-sum vision. From such a viewpoint it may seem a natural step to establish progressively tighter contractual relationships for succeeding projects so that potential liabilities to a protected party become minimised. As the consequences continue to be unsuccessful – other than for those who profit from increased consequential litigation – further tightening is imposed so that the project is literally strangled at birth. These notions of sharply separated responsibilities should be confined to simple readily defined transactions not subject to uncertainties of any importance beyond the control of the contracting party, e.g. the purchase of a fully specified pair of socks.

Where the project managers are not competent to engage in 'conducting' the project, they fall readily into the zero-sum mode of operation in the company of those who lack understanding of engineering principles and have turned to 'management' in compensation, who follow the trail, without comprehension of the cost of the consequences. Once the disease takes hold, often from an uncomprehending and project-innocent board of management of a prospective Owner, it is highly contagious, leading to a bureaucratic project control structure, superficially efficient since all costly changes and rectifications can be blamed on somebody else.

In the construction industry the market depends upon the Owner or Client who determines not only the nature of the product but the terms under which it is to be provided. Whether they are destined to get what they need, expressed in terms of performance and value for money, depends on the skill with which their requirements are translated into appropriate terms for the provision, through an appropriate procurement process and subsequent management. Too often, with a predominantly legal and financial background, the Owners have no understanding of the criteria for success or for avoiding the obstructions to success, imposing over-rigid terms and allocation of risk in the face of uncertainties which are in part governed by their own transactional and adversarial behaviour. Where a project is founded on a combative base there is no scope for innovation so that risks occur in the absence of protective mechanisms.

An immediate consequence is that of project fragmentation. Each element that should contribute to the holistic approach to the project is commissioned, usually on the basis of least cost, without consideration of the likely evolution of other elements, as has been outlined in Chapter 2. Within the Owner's office the only measure of efficiency relates to the delivery of each commissioned fragment to the specified date and content, but not to quality or to the contribution to the project as a whole (the holistic concept). In this environment, contracts tend to be prepared to standard models, each professional contributor to the process is designated a 'contractor' to emphasise the constricted nature of the 'deliverables'. The management jargon is fulfilled, each successive piece is inserted in the paperwork and the ultimate cost mounts insidiously without accusatory evidence. The zero-sum mentality prevails and no competence exists within the client's structure to compare the overall value for money for such a 'crazy-pavement' form of project against a truly optimised project.

Moreover, a form of Gresham's Law develops, with the inferior driving out the superior, where professionals are engaged in this manner. Those with least capability for insight of the potential problems will not foresee the complexities of the commission to be undertaken professionally, i.e. beyond the commercial limits, and their involvement will be favoured on lowest cost, often sharing with the Owner – who knows no better – the ignorance that he is thereby short-changed. A valuable capability of the professional is to advise on the terms of reference appropriate to enabling the Owner to receive the results to be expected from a well conceived procurement process, which may be of great value and relevance to the project. When the professional is treated on a commercial basis, this procedure is eliminated. The right or vital answers cannot be expected from the wrong questions. Management jargon overwhelms; leadership is lacking.

Currently, there is a dividing of the ways. While many organisations become yet more entrenched in the practice of project fragmentation, the enlightened few perceive that release from the history of poor project performance springs from a totally different course. This entails acceptance of the logic of *design* (as defined in Chapter 2), managing the project to permit greatest benefit from the combined skills from those who contribute, the art of 'management by *design*', even though few – yet – use the terminology of this book!

7.4 The functions of project management

The essence of successful project management derives from management of the *design* process (Section 2.1). This entails the organisation and control of the resources appropriate to the project, establishing a coordinated and methodical convergence towards the definition and attainment of the project targets, rooting out risk and rooting for beneficial innovation, by synthesis of the contributions to be expected from the parties in the *design* process. Supporting measures include the consideration of the formal processes for decision-taking, for change, for assembling resources and for their reimbursement, for reporting on the project and for overall control of quality.

Good project management should flow from appropriate project planning, discussed in Chapter 3. The planning process will have attempted to achieve the optimal balance between costs of construction of the project, expressed in terms of value for money, and its operational performance. New information or development may require this equation to be periodically re-examined, always mindful of the high costs of any attempt at modifying what has already been built or placed on order.

The principles of Quality Assurance (QA) have a part to play in establishing procedures for quality control and in ensuring that these procedures are fulfilled. The main limitation of QA is that it is concerned with foreseen quality issues whereas tunnelling essentially provides occasions for unforeseen features and for characteristics which do not lend themselves to the simple yes/no options of QA. Moreover QA, with its own viral form of Parkinson's Law, tends to expand into areas far better left for engineering judgment and, uneconomically, into procedures and 'house style' unrelated to quality and safety, attended as they then will be by rigorous rules for controlling change and ensuring compliance. As a consequence, much time and energy is diverted to negotiating escape routes from non-compliances with QA, where simple and valid engineering solutions have to be ignored because they do not fit requirements to follow procedures of apparent Byzantine complexity lacking technical justification.

In applying principles of QA, emphasis should always be given to exercising the powers of observation on the look-out for precisely those features which are not included in the QA procedures. Otherwise QA tends to provide excuses for failings in vigilance, and QA is no substitute for understanding in depth the nature of the engineering uncertainties and the features which may give advance warning of trouble. A QA engineer may be appointed to ensure compliance with QA and, if so, the project management needs to establish the role within the overall objectives for the success of the project, including assurance as to the availability of records in such a form as to be of value in identifying and resolving potential problems.

Quality assurance has in certain instances been confined to what are defined as 'permanent works', i.e. excluding a Contractor's Temporary Works. In tunnelling, this is a very artificial division. Whatever may be the precise arrangements for working arrangements, which may well be related to the Industrial Regulations of the country concerned, QA must include all aspects vital to the permanent works. For the tunnel elements of the Øresund Link (Hentschel 1997), for example, the bulkheads at each end of the element were supported against temporary horizontal reinforced concrete edge beams attached to the base of the roof and to the top of the floor of the element. As temporary works, these steps escaped QA which was being relied upon to ensure performance of the Contractor's procedures for ensuring compliance of workmanship with design. As a consequence, a bulkhead failed during sinking, fortunately without loss of life or gross damage. This might have been a far more serious mishap. Every finished tunnel emerges as a result of metamorphosis through several stages of construction. The success of each is vital to the project. It makes no sense if QA is inattentive to these vital intermediate phases for the formalistic reason that they do not constitute the 'permanent work'.

A practice has developed for replacing independent inspection of the construction works by 'self-certification' to the effect that certain features specified by QA have been undertaken. The expectation has evidently been that such a practice confers comparable security. It only achieves this purpose when self-certification is seen as part of a professional undertaking and not as one performed against conflicting commercial pressures, concerned with the formalities of QA and no more. During the course of a project, features of quality specified from the outset may benefit from being revisited, as it becomes clearer as to which features are vital, which cosmetic in nature. Tunnelling, particularly those of its features to be hidden by subsequent construction (often the most vital elements), entails operations which require experience and judgement to establish whether they have been undertaken in such a manner as to fulfil the expectations of the designer. Those features, most essential and possibly arcane, need the resource of a knowledgeable engineer capable of taking a detached and responsible view, who needs to be aware as to which is vital to fulfil the intentions of the designer. Where there are departures from these intentions, the inspecting engineer needs to have power to order immediate action, over-riding apparent conflict from commercial pressures.

Emphasis is repeated throughout this book (see particularly Chapter 2) on the 'orchestration' of the tunnelling project with the manager seen as the 'conductor', ensuring the integration of the principal participants. A project does not start from a *tabula rasa* but draws upon much experience, direct and indirect, from similar or at least comparable projects, as illustrated by the helix of Figure 7.1. Success in management largely depends on the ability, directly or by proxy, to draw upon, distill and blend such experience, learning the lessons of failure as well as success. Features of uncertainty are identified in this process (Section 2.1) providing the basis for risk analysis and control as the project is defined, designed, constructed and operated. A skill, expected particularly to be displayed by the project manager, is that of transmutation of experience, recognising that circumstances between projects – or even between parts of projects – are different, sometimes subtly different. For example, where a particular incident has occurred in a tunnel, it is necessary to understand the contributory factors, how these combined and the lessons, in particular and for wider application, to be learned.

The development of the successful project may be visualised as a convergent helix, illustrating the interactive nature of the process and the constant communications between participants in the *design* process towards the optimal goal.

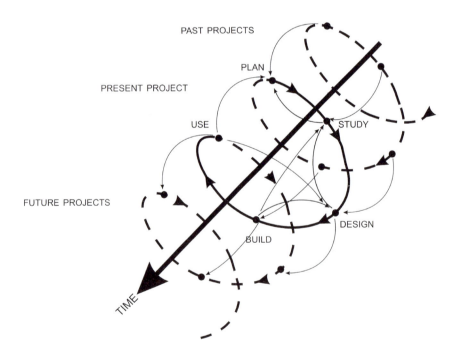

Figure 7.1 Tunnelling: the iterative nature of learning.

The dominant questions for a tunnel – as for any other project – are too readily obscured by detail but must remain exposed to view and to critical development:

Why? – the purpose of the project and how success is to be measured.
What? – the nature of the project to fulfil the purpose.
How? – the means and the method of construction.

These questions need to dominate the construction planning process. They cannot be approached sequentially – to be more correct they frequently are and those guilty reap the consequences – as explained by Chapter 3. As the construction process and the means of construction, particularly special plant, become more complex and specialised, so does 'How?' occupy an increasingly important role in project planning. 'How?' in tunnelling is a dominant feature, controlling not only the economics of choice of scheme but possibly the more fundamental issue of feasibility. What, for example, is the greatest practical depth at which a particular form of watertight tunnel may be built in water-bearing ground?

An art of balance concerns the gradual imposition of restraints on the options for 'What?' and 'How?'. The elimination of one option will require a certain amount of information in order to establish the superiority of another, within the uncertainties of knowledge, superiority in terms of expected life cost, suitability in operation, time of construction, control of risk. Where comparisons are made between options, it may be possible to accelerate the process (Figure 3.4) by eliminating common features of uncertainty, in a manner comparable to that described in Section 3.1.3 for focusing upon optimal planning scenarios, never forgetting in so doing that risk of a particular scheme depends on specific susceptibilities of the scheme to a particular set of uncertainties. Indeed, the uncertainties may group themselves differently in their combined impact on a particular scheme, by comparison with another (Figure 2.4). The greater the number of options retained at any stage of a project, the greater the costs in assembling the data necessary for the next sequential phase of refinement. One essential feature for success is the open sharing of all information relevant to the *design* process, the feature designated as 'transparency'. As a practical measure, common computer programs should be used between the Parties for recording progress and as design packages for producing drawings, thereby helping to avoid conflict between the requirements of different participants and of relating permanent works to temporary works. It is a paradox that where, on the contrary, in disregard for good practice, attempts have been made to throw risk disproportionately onto a single party, the withholding of any information germane to the assessment of such risk may in litigation and at great cost thwart the attempt to do so. In consequence the Owner loses all: the possibility of a successful project; the self-inflicted costs of litigation.

The equation between time and money needs to be carefully studied for a project. For the Client as operator, the timing of investment needs to balance the cadence of construction costs (unless these are to be accompanied by financing proposals), with revenue, actual or notional, derived from the expected degree of usage after commissioning. To a certain degree, the cost of construction can be estimated in relation to the rate of working for a given size of project, taking account of costs of mobilisation and demobilisation, special plant needs, learning curves and so forth. Tunnelling is subject to particular constraints as a result of the linear nature of the operation with limited access. The number of faces may not be readily varied for an underwater or urban tunnel although this may be an option on occasion. The frequency of working-shafts for many of the ancient tunnels described in Chapter 1 was governed by a conscious, and project-specific, objective for optimal balance of time and cost.

Throughout, the objective must be to modify previous planning so as to take best advantage of new information or interpretation of data old and new. This process requires fullest cooperation of all concerned, encouraged by a shared interest in the stated criteria for judging the success of the project. A newly encountered problem in construction, for example, may best be countered by a revision in layout or project design.

7.5 Principles of project management

The management community contains obscurantist elements, typified by those management consultants who find the need for constant invention of this year's modish neologism or acronym which, on analysis, reveals little more than the simplistic approach of its agent. The object appears largely to envelop the art in a mystique to encourage the belief that guidance is essential from the true believers, the acronymic guru. There are however a number of principles for success which may be stated briefly:

1. Overall competence of the team, appropriately assembled under good leadership as the project develops. The possible need for a 'surrogate operator' to inject operational competence where this is lacking in the Client has been discussed in Chapter 3.
2. Methodical approach to project definition, which starts from the identification of the 'business case', i.e. the statement of the objectives and assumptions which justify proceeding with the development.
3. Appropriate resources, which need to be available to take the project through its preliminary stages without hiatus. Of course, the precise needs for financing, and the optimal means, develop synergistically with the definition of the project.

4. A shared interest in success between all those contributing to the overall *design* process. Means to achieve the objectives will be project-specific and are discussed further in Section 7.7.

5. Following from the above, there must be continuity in responsibility as the project develops. For a road tunnel, for example, Figure 7.2 summarises the technical input requirements as the project develops. The 'centroid' of major input changes as the project passes through different phases but the guiding philosophy must be sustained. For example, it may be necessary to reconsider an earlier decision for which purpose the context and influencing factors of this decision must be accessible and revisited.

The shared interest in success, already emphasised in Section 7.3, requires particular emphasis. The author has heard more than one management consultant discuss project success in terms of his company's profit, regardless of the abandonment of the project in question.

Following from these principles, synthesis must be assured of the activities of the several contributions, with interactive iterative progression with time. Risk analysis and control will be central to all decisions, an operation which requires constant communication and interaction between participants.

A fundamental conclusion which follows from application of such principles is that relationships between the parties in the form of contracts and agreements must conform to the requirements for project optimisation and not be imposed in cavalier manner in conformity with an Owner's standard, imposed, as too often occurs, regardless of the nature of the project. Furthermore, what appears frequently to be overlooked is that, where project risk is imposed upon the contractor, project direction is necessarily transferred in association. What then follows is the ludicrous – were it not also highly risk-prone – masquerade of nominal (hands-off) project direction by the Owner and advisors (in fact, largely administration of payment) while the real project management rests on the shoulders of the Contractor.

7.6 Project management in practice

As stressed throughout this book, management needs to be deeply embedded in the *design* process. There is no place for 'hands-off' management which only obstructs the essential interactive elements of *design*. A special skill is that of the anticipation of approaching problems, as opposed to the far more expensive practice of reacting on their arrival. Where there is common interest in the overall success of the project, there is a likelihood that reports on progress and on minor incidents, that may together presage some greater problem, will be factual, informative and

STAGE OF DEVELOPMENT OF PROJECT

DISCIPLINE	Conceptual Planning	Conceptual Design	Conceptual Design	Construction	Fitting out	Operation
CIVIL ENGINEER (ROAD TUNNELS)	*	*	*	*	*	o
CIVIL ENGINEER (ROADS)	*	*	*	o	*	*
TRANSPORT ECONOMIST	*	o	–	–	–	o
PHYSICAL PLANNER	o	*	o	o	o	o
ARCHITECT	o	*	*	o	o	o
GEOLOGIST	o	*	o	o	–	–
GEOTECHNICAL ENGINEER	o	*	*	*	–	–
LIGHTING ENGINEER	o	o	*	o	*	o
VENTILATION ENGINEER	o	*	*	o	o	o
FIRE OFFICER	o	o	o	o	o	*
EMERGENCY SERVICE OFFICER	o	o	o	o	o	*
ELECTRONICS ENGINEER	–	o	*	o	*	*

1. THE CONCEPT AND EXECUTION OF A ROAD TUNNEL REQUIRES THE ASSEMBLY OF EXPERTISE ACROSS A WIDE RANGE OF DISCIPLINES AND EXPERIENCE. THE GREATEST SKILL IS THAT OF THE SYNTHESIS OF THIS KNOWLEDGE AND EXPERIENCE.

2. IN TABLE
   * REPRESENTS EXPERTISE, AT A LEVEL APPROPRIATE TO THE PROJECT.
   o REPRESENTS GENERAL UNDERSTANDING OF THE AREA OF EXPERTISE.

Figure 7.2 Principal technical contributors to a road tunnel.

objective, and not protective and over-confident. Where the interests of the participants do not coincide, e.g. as the result of the arbitrary or wilfully inappropriate allocation of costs and risks, the first interest will be to establish liability and reporting will tend to be partisan. One such example occurs when conditions are a little more unfavourable than an optimistic interpretation of the initial geological data. Contractors have been known, in such circumstances, deliberately to exaggerate difficulties of construction to establish a case of 'unforeseeable conditions' where none exists, at the expense of maintaining progress. At a time when the greatest degree of cooperation is demanded to forestall what may develop to a disaster, the lawyers will advise: 'Do nothing. Sign nothing. Speak to nobody. Sit absolutely still'. This is the nature of a 'brittle' contract which, by the way it was constituted, made no provision for the flexibility of approach necessary to find solutions to new problems. Often the Owner, who has willed such a situation in ignorance of its consequences, is the principal sufferer. The Contract contained no built-in machinery for resolution of problems which required cooperation. The trench warfare – covered trench warfare perhaps – which ensues is a totally unproductive invasion of the valuable time of skilled engineers.

The term 'holistic' is possibly overworked but it is a vital attribute to avoiding the misallocation of risk. An holistic approach implies that all contributory factors will be addressed towards a solution. Throughout the evolving project an holistic approach will help to find opportunities for new solutions to perceived problems which could not evolve from a partisan or piecemeal approach. Even during the preliminary stage of a project there may be a need for rapid decisions involving several interests. For example, a site investigation reveals information unexpected to a degree to require revision of the ground model (Appendix 5F) and to consequent rethinking of project strategy. In this way, problems are solved in such a manner as to reduce the sum of the combined exposures of the Parties, to mutual benefit in avoiding the 'zero-sum' constraint.

During construction, the needs for rapid decisions are obvious. In general, the nature of possible problems should have been considered and noted, together with the nature of the path to be followed to ensure prompt deliberate counter-action, recognising the several aspects that need to be satisfied in solving the problem. An important benefit of the observational approach (Section 2.6) is that a systematic procedural system is already in place when required (Figure 2.9). Where a particular form of hazard is perceived, a well-defined contingency plan will need to have been evolved in order to contain the risk to within acceptable levels, to be brought into effect without delay in mobilising resources. This was, for example, the situation during construction of the Cargo Tunnel at Heathrow Airport, described in Section 5.2.3.

The observational approach may play a prominent part in the apprehension of features that may lead to a risk. There is evidence of confusion as to the scope of the Observational Method which, in the nomenclature of this book, is known as 'Observational Design' (OD). Observation may be undertaken solely for the sake of obtaining information on performance or on behaviour of the ground or tunnel. When observation entails monitoring (Figure 6.5), i.e. the ascertaining of performance with a possible view to consequent action, this constitutes OD, whether the need for such action may be seen as exceptional (Case 1) or whether the design has explicitly accepted the occasional or more frequent need for supplementary work (Case 2), on a basis such as that discussed in Section 2.6. It is simply a matter that the probability of the need for supplementary work under Case 2 is greater, possibly orders of magnitude greater, than for Case 1. Case 1 expects confirmation of the adequacy of the initial provision but needs nevertheless to include contingency planning for the unexpected, but possible, departure from expectation.

7.7 The team and the contract

Recent evolution of contract practices for tunnelling has been generally disappointing. In the United Kingdom, to take one example, by 1978 there was a particularly promising background and tradition. The minimum criteria for success were understood as: enlistment of competence, holistic *design*, continuity in direction – all were features of the tradition of the contract between Employer and Contractor directed by the Engineer, whose considerable powers correlated with duties.

Under the Institution of Civil Engineers Conditions of Contract (ICE Conditions) up to the 6th Edition, the Engineer is the independent administrator of the Contract. This also applies to the international counterpart, FIDIC (Fédération Internationale des Ingénieurs Conseils) Conditions up to the 4th Edition. The functions of the Engineer also include services as Agent to the Employer as planner and designer of the Works, possibly in addition as general advisor. The Engineer needs at all times to be clear as to which function is appropriate to each act of participation. The wise Employer appreciated that the confidence of the Contractor in the objectivity of a respected Engineer as Contract administrator much reduced the need for hidden provisions in pricing for uncertainty. Provided the Contract was constructed on equitable lines, the Engineer would recognise an unforeseeable event, as qualified by the Contract, with the responsibility of considering all aspects, including possible modification to the design of the Works (by the Engineer as Employer's agent) to achieve the most satisfactory solution. The Engineer was also responsive to suggestions from the Contractor to modifications to the scheme of work which had no detriment to the project. The system worked well for readily definable contracts, when:

1. the Engineer was fully competent to undertake the range of tasks entrusted to him;
2. the independence of the Engineer was respected, without unreasonable constraint (such as a requirement for the Employer's prior agreement to any ruling by the Engineer);
3. the Employer did not make a 'reflex' recourse to his lawyer when a cause for variation had been unforeseen by the Engineer as well as the Contractor.

In fact, much of the criticism of the system originated from lawyers who grudged the extent of the powers of the Engineer. Essentially the system depended on the Engineer acting proficiently and loyally (as Agent) in the interests of the Employer and as a professional of independence and integrity as Contract administrator, a feature of a profession at odds with the more evident self-interest of commerce. The Engineer, under the system, was the 'Conductor' as described in Section 2.1 who could thereby bring the virtues of operating the *design* process to the project.

When operated by a knowledgeable Employer (as to the objectives), an experienced Engineer and a Contractor appointed on merit, the system worked well for the time and was not, as later critics imply who never saw the system operating effectively, a confrontational system. Confrontation only developed when expansion of work attracted inexperienced Engineers, appointed in competition predominantly on cost, whose powers were fettered with regard to the degree of independence from the Client (Employer), with invasion of the relationship of trust between Employer and Engineer by the intervention of the lawyer. The enlightened Engineer had previously accepted large responsibility for the broad area of *design*, with opportunity to ensure that the terms of appointment from the Employer adequately expressed project objectives, supplemented by constant discussion during project definition to refine requirements against likely costs, and prepared to modify project design in response to unexpected circumstances.

With the increasing importance in tunnelling of the means and methods of construction (process design), procedures were occasionally modified to include specialist advice from contractors or specialist suppliers as the project developed, where the Engineer recognised, as a result of new developments, the desirability of resort to know-how beyond in-house experience. Nevertheless, the system began to creak for large complex projects where greater unification of effort was desirable between the 'players' and where greater flexibility was required to determine the scheme of work, including possible acceleration, best to suit the challenges of uncertainty in the interest of the project in order to respond to the unforeseen. Fundamentally, concerns for operation of complex projects assumed greater importance so there was increasing need for the Owner, as future operator, to take an active part in the development of definition of the project – a process often continuing through construction to commissioning.

The engineering profession remained largely complacent as the system deteriorated, possibly distracted by the high work load at the time. It may even be suggested that, in the short term, the incompetent within the profession, whose failings contributed to the later criticisms of the system, were profiting from the system's defects. This was the occasion for the lawyers, aided by other ancillaries, to pounce. The positive move should have been towards a fundamental rethinking of the project scene, drawing upon the professional integrity of the engineers and others involved in all aspects of construction. Instead, it appeared modish to encourage a drift towards increasingly partisan and complex contract documents (the disease was particularly rampant in the English-speaking world – one large project in Canada had such a complex set of contract documents that only one man, a lawyer, was said to have read them all) with increasingly defensive postures of the Employer reflected in the attitudes of the Contractor, forced to compete at sub-economic rates for a pessimistic outcome, while the Engineer was compressed into a subservient role with overall duties fragmented and diminished. The relationships became fragmented and confined to commercial bases. The result, as has been widely recognised but too rarely correctly diagnosed, is that projects, lacking the necessary cooperation across the component parts, are well below optimal achievement. (The 30% suggested economy proposed by the Latham Report (Latham 1994) is readily achievable in tunnelling, by comparison with the out-turn of recent 'fragmented' projects by the adoption of good practice, i.e. management by *design*, the title of this book.) The consequent level of litigation, a direct result of the deliberately confrontational contracts imposed by the influence of the lawyers, should immediately recede. In 1978 a constructional lawyer in Britain was a rare breed. A result of the restoration of good practice is that there is not a lawyer in sight in discussions between the Parties, the Parties having common interests in contributing to objectives as opposed to contractual opposition. In 1978, as described in Chapter 2, CIRIA published *Tunnelling – Improved Contract Practices* (CIRIA 1978) which built consideration of risk-sharing and the means of reducing uncertainties between contracting parties onto traditional contracting practices, directly opposed to the trend of retreat into defensive postures.

The organisation of a tunnelling project may be visualised as a series of systems which interlock or which may be wholly contained one within another. The system represents the *design* process in action. The contractual relationships need to be so arranged as to be compatible with the operation of each desirable system or sub-system, advancing and not impeding the process of project optimisation. The consequences of such an approach lead to a choice of forms of relationship expressed in formal terms suitable for the circumstances. As described below, the particular project may, for example, be best served by a form of 'partnering' at one extreme or by a Lump Sum Contract at the other, with several intermediate choices.

A general requirement of good practice is that all information known to the Employer or Engineer, with a bearing on the responsibilities placed on a Contractor, should be made readily available. This requirement will include all relevant geological and geotechnical data as described in Chapter 3. Two particular features require further clarification and emphasis:

1. Interpretation of data against a wider background and against experience of comparable projects may be vital to the informed use of the raw data in assessing problems of construction. Often such work of interpretation has been undertaken in depth and over a considerable time. A Contractor should not be expected to have repeated such work within a brief competitive tendering period. There may well be a formal disclaimer by the Owner concerning liability for such interpretation. Where the interpretation leaves open features of appreciable impact on the system and cost of tunnelling, use should be made of the principle of stating 'Reference Conditions' (CIRIA 1978, para 4.7).
2. In the public interest, all information derived from ground investigation should be accessible for those planning new underground works or modifications to existing works. Judgment is needed, where the existence of such data is known or suspected, by the engineer responsible for a new project, as to whether access should be requested to the data.

A first consideration in project organisation concerns the nature of the operation. A tunnel seen as a conveyor of stated capacity between points A and B may be designed without further recourse to its performance specification beyond efficient functioning over a stated period of years. Underground construction for more complex purposes, e.g. gas storage or transport tunnels, will require constant iterations between operational desiderata and constructional options, with the understanding that the economic objective is to minimise the total of the capitalised costs of construction and of operation. In the latter instance, the operator will play a key part in the design management and in the discussions that will follow the identification of any unforeseen feature or new opportunity.

Figure 7.3 illustrates an organisational system for tunnelling for a specific purpose of linking A and B through ground familiar to tunnellers which experience has shown to be without geological surprises, a feature to be checked for the specific project. Features of functional planning are

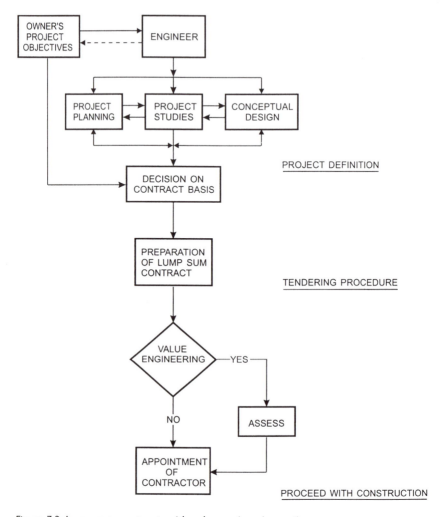

Figure 7.3 Lump sum contract, with value engineering option.

translated, by a knowledgeable engineer, in conjunction with appropriate studies, into the project design of a familiar nature. This may then, for the construction phase, constitute the input for the system of construction and monitoring, since the proposed construction methods do not impinge on the project design requirements. For a very simple instance, with no expectation of external variation, a Lump Sum Contract may be considered. More generally, the familiar circumstances may be favourable for an element of innovation offering benefits, requiring an element of interaction between project design and construction design. Here the preference may be for a Lump Sum with an alternative which offers shared benefit to Contractor and Employer for savings for the alternative measured below the Lump Sum value. This is a form of Target Contract which does not permit increase beyond the Target, a specific form of the more general expression: 'Value Engineering'. Where used in the USA, a valid criticism is that the Lump Sum Contract has often been based on highly conservative schemes which readily lend themselves to savings, occasioning greater ultimate cost to the Client than would arise from a better engineered project in the first instance. This has often stemmed from the separation between the design and the construction phase in North America which has resulted in many designers having limited experience of construction (and constructors of design). This traditional separation in the USA coupled with the dominance of the construction lawyer represent features which have held back innovation. The project planning and associated studies have to be conceived as adequate for any option that may reasonably be proposed or there will not be a valid basis on which to assess the merits of the alternative. There is otherwise considerable risk of problems arising from lack of continuity between the conceptual thinking and its development by the contractor.

Target Contracts (Figure 7.4), as encouraged for example by the Institution of Chemical Engineers (1992) for Process Plants, are held by many to reduce the antagonistic attitudes believed to be inherent to the form of contract based on Measured Works. Target Contracts are said to encourage a common interest between the parties to the Contract in reducing cost. Provided the project costs remain within the target figure, the benefits of efficiency are indeed shared. The problem, particularly in tunnelling, arises from the unexpected when the target value is exceeded and when it is then in the Contractor's interest to renegotiate the Target Value in order to provide a new base line for the calculation of fee and thus to recover loss of profit which would otherwise occur. As illustrated (Figure 7.5), this feature may have far greater incidence on the total cost of a project than would alternative arrangements for variation of a measured contract, following the pattern of ICE Conditions or FIDIC Conditions, as described above. Variation for unforeseeable conditions of the ground may in either instance require Reference Conditions (see above) to provide a base-line for negotiation.

A real problem arises where, as often accompanies a Target Contract, the Contractor has latitude for the choice of the means of construction, a choice which may have influenced appreciably the exposure to risk (Section 2.1). Certain Employers, or their advisers, deliberately choose generous Target Values with the expectation that this should be readily achieved by an efficient contractor, to eliminate the eventuality of overrun but, if so, the claimed benefits of a Target Contract are materially diminished and the Employer enjoys spurious, self-congratulatory credit

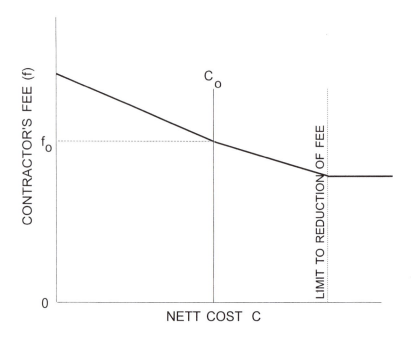

$$P(\text{PERCENTAGE FEE}) = f/C, \text{ SO } p_0 = f_0/C_0$$
IF CONTRACTOR BEARS n% OF INCREASE IN NETT COST, ΔC
$$P = f_0 - n\,\Delta C/C. \quad \text{FOR COST SAVING,} \quad p = f_0 + n\,\Delta C/C$$
(n MAY BE 50% FOR SAVING, 30% FOR INCREASE)

Figure 7.4 Target contract.

for the apparent virtue of controlling project cost. In the absence of valid bench-marking, the generosity of a specific Target Contract cannot be estimated. This is a circumstance in which the Engineer's estimate, based on good practice in the circumstances of the project, with stated provisions for uncertainty, can provide a valid basis for comparison. The more effective control is to set a realistic target and to undertake good quality engineering examination of a Contractor's proposals and method statement in order to establish that foreseeable geological surprises are unlikely to introduce problems of a nature to support variation of the Target Value.

The Author can instance major projects of his own experience, e.g. the Clyde Tunnel (Morgan *et al.*1965), Potters Bar Tunnels (Terris and Morgan 1961), Heathrow Cargo Tunnel (Muir Wood and Gibb 1971) and others, including for example the Orange-Fish Tunnel, where a remeasured

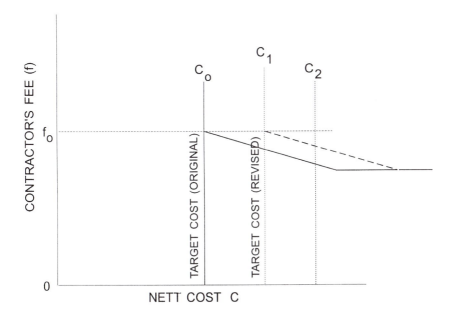

REVISION OF TARGET COST FROM C_0 TO C_1, FOR AN OVER-RUN
IN COST TO C_2, INCREASES COST + FEE BY

$$(C_1 - C_0)\left[(1 + f_0) + n\right]$$

Figure 7.5 Effect of revision of target cost.

contract administered by an independent Engineer has provided a project
meeting the requirements of the Owner, without dispute and achieved
within time and budget. The first of these projects encountered a number
of different forms of geological hazard, the problems requiring design of
specific tailor-made solutions as each was more specifically defined as the
work advanced. The second and third projects were innovative in basic
design and required the degree of coordination between design and
construction compatible with such an arrangement of Owner, Engineer
and Contractor. Where the Engineer sees high benefit for the project from
innovation with acceptable concomitant risk, combined with fully meeting
the specified operational requirements, as for the Heathrow Cargo Tunnel,
such a system is particularly appropriate. The Author insisted for this
project that ICE Conditions should be used in order to provide powers
of the Engineer adequate to establish the essential coordination between
design and construction (as opposed to a Government model CCC Works 1
preferred by the Owner). The innovative design, saving at least 50% of

the project cost to the Owner, could not have been accepted for a Lump Sum or Target Contract, and certainly not for a project in which the several contributions were fragmented and contracted each at least cost. The notion of project interacting with process is fundamental.

Where the ground is uncertain, while efforts may be made to ensure that the direct expenditure on special expedients (Section 6.4) is neutral to the Contractor's profit, the overall balance is difficult to achieve since the consequential easing of the Contractor's task, e.g. by reducing inflow of water below an acceptable amount, will probably not figure in such a calculation. This, once again, is a situation in which a remeasured contract may be the fairest course, in the interest of both the Owner and the Contractor.

In 1993, the Institution of Civil Engineers (1993) issued a new basis for construction contracts, the New Engineering Contract (NEC), followed in 1994 by a second edition with a minor amendment titled *The NEC Engineering and Construction Contract*. These documents contain many innovative features of potential benefit. The commissioning of major projects in Britain at that time was being undertaken without adequate contribution by engineers with a grounding in the criteria for project success. The documents represent considerable departures from traditional practice and should have been accompanied by a comprehensive guide to the Owners as how to apply the spirit of the scheme to the best advantage of a particular type of project. It needs to be recalled that project fragmentation – with each element, of planning, of design, of site investigation and its supervision, of construction obtained in a competitive manner – was the prevailing commercially-led scheme for project procurement in Britain. Certain features of the NEC, including the sub-division of the traditional functions of the Engineer into designer (a function indicated as complete by the time of tender), Supervisor and Project Manager, appeared to encourage the practice of fragmentation.

The NEC provides five different bases for payment to a Contractor:

A – Priced contract with activity schedule;
B – Priced contract with Bill of Quantities;
C and D – Target Contract (with payment as A or B above);
E – Lump Sum Contract.

An objective is stated that the 'NEC is intended to provide an up-to-date method for employer, designer, contractors and project managers to work collaboratively and to achieve their own objectives for their work more consistently. . . .' intended to lead towards much reduced risk of cost and time over-runs. Changed requirements are recognised through 'compensation events'.

The treatment of ground conditions, so vital for tunnelling, however falls short of good practice. For tunnelling works, a few simple 'boundary conditions' are suggested as confining the Contractor's exposure for uncertainty about the ground. There is no indication of appreciation of the subtlety of such criteria in relation to methods of working, i.e. the relative tolerance or adaptability of the system (Section 6.3). Nor is there clarity concerning the provision of data. The guidance note with the NEC documents advises (p. 27) that only factual information should be provided to tenderers – with the hint that some Employers may go rather further, while p. 59 advises the provision of interpretation of factual data by a specialist. A major issue of potential uncertainty then derives from the geological risk imposed on the Contractor. Clause 60.1 limits the Contractor's liability to 'physical conditions which. ... he could not reasonably have expected to be a significant probability'. If the information is inconsistent, the Contractor is assumed to rely upon the less adverse (in the 1994 version, the more favourable). This is unhelpful for the Contractor left to choose between an 'intolerant' system potentially more economic than a more 'tolerant' system, for which a more cooperative approach between the Parties is to be preferred.

The NEC, with fairly minor variations, in respect of such issues, is capable of acceptance for tunnelling provided it is set into a broader framework which will establish:

- full coordination across all preliminary activities of planning, design and supporting studies, including site investigation, directed by the Employer with the Engineer or by the Engineer on behalf of the Employer;
- development of a practical scheme (or alternative schemes) for construction pre-tender in order to establish adequacy of provisions and constraints of the Contract;
- provision for design continuity between pre-Contract and the construction phase, in order to encourage innovation, reduce risk arising from uncertainty and to develop professional relationships between the Parties;
- full continuity of risk analysis and control throughout;
- a system for full integration of operational objectives with construction, with continued guidance to the Contractor and appropriate levels of detailed provision of method statements in return.

Providing such a basis is clear from the outset, such a scheme may be devised to be compatible with payment on activity schedules of Scheme A of the NEC. Provision for beneficial modifications need to be made so that interest is at all times focused on the requirements for the Project as the over-riding concern of all involved. The essence entails the broadest analysis of the way in which contributions from all directions may find optimal solutions to problems encountered during construction.

A common occurrence for tunnelling is where the general nature of the ground, and its variability, may be predicted reasonably well, but only roughly the extent to which the ground may be 'zoned' and the precise position of each zone not at all. In these circumstances, type design for tunnel support may be provided for each zone and serve as the basis for payment. So far as practicable, an objective basis is provided for the zoning, which may be found to require some degree of revision in the light of experience as construction proceeds.

The Øresund Link discussed in Section 7.2 provides a good example of project management based on a variant of FIDIC Type Contract administered by the Owner, dependent on considerable contribution from Consulting Engineers, with payment based on achievement of 'milestones' supplemented by Variation Orders on the basis of changed conditions. The principle of risk sharing is accepted, including a set of Reference Conditions, with continuity across project and construction design, also in the management of the interfaces between principal Contractors which would be impossible in a situation of confrontation. Management is by engineers with lawyers confined to the proper functions of lawyers.

As the different aspects of the project interact more comprehensively, so does the case become stronger for a form of partnering between those contributors responsible for the elements concerned. Schemes of PFI (Private Finance Initiative) usually involving DBO (Design, Build and Operate) provide opportunity for unified engineering throughout the project. The potential for 'internal partnering' in a DBO context remains underdeveloped. Engineers, previously confined to particular roles in a confrontational setting, tend to remain on the defensive within their own perceived area's of responsibility, while excessive powers of management remain in the hands of those who do not perceive the merits of interaction and are in consequence blind to the benefits of operating the *design* process. People who have traditionally accepted responsibility for a confined aspect of a project appear to feel insecure when they are expected to share responsibilities alongside the consequent benefits of managing a continuity across the several inter-related functions. As a result, the worst features of traditional posturing may be encountered, standards further depressed by the overall control having a commercial rather than professional motivation, to the detriment of the project. It is essential that education, training and operation of project 'transparency', i.e. the sharing of motivational objectives, should contribute towards eliminating such divisive attitudes, in combination with rewards related to success of the project overall. Partnering offers benefits in providing scope for full optimisation, which is not properly exploited if the elements of design and construction are kept at arms length from each other and from the operation of the completed project.

Another feature of the DBO project concerns the remarkable number of consultants employed on behalf of different agencies to undertake essentially QA type duties for each separately. Apart from the proliferation of bodies making for a complex bureaucracy, the enemy of good engineering, the result may not achieve the objectives. The purpose of each is to 'sign off' the discharge of a particular function. The absence of integration of functions may lead to unforeseen gaps in the provision. For instance, the overall standard of safety of a road tunnel in a fire depends on the traffic, the control system, the layout, fire-fighting and fire-escape provisions. Assurance of adequate standards requires all such features to be considered as a whole, not each in isolation from the others.

Partnering is a term which was originally coined in the petroleum, processing and power generation industries in the United States. The Construction Industry Institution (CII) Task Force on Partnering provided a definition in 1987: 'Partnering is a long-term commitment between two or more organisations for achieving specific business objectives by maximising the effectiveness of each participant's resources. The relationship is based on trust, dedication to common goals, and an understanding of each other's individual expectations and values. Expected benefits include improved efficiency and cost effectiveness, increased opportunities for innovation and continuous improvement of quality products and services.'

While partnering remains predominantly a practice between large international clients and their principal contractors engaged in serial projects, the potential scope is now recognised as considerably wider, where the processes of construction may have particular influence on the life costs of the project and where operational expectations dominate the criteria of design. Cooperation between participants will in consequence promise, and achieve, results superior to traditional, or to more recent adversarial, practices. The contract is rewritten in these circumstances as a statement of objectives and as providing a basis for reward related to success of the project overall, treating the participants as professionals, eliminating the dominance of the legal profession and providing in consequence escape from the 'zero-sum' game (Section 7.2), while recognising the community of objective in overcoming the consequences of unforeseen circumstances.

For the time being, partnering for major underground projects will doubtless be confined to those Clients who do not depend excessively on finance by commercial banks or dominance by accountancy, with its concomitant distortions of risk allocation, legalistic relationships and management separated from engineering. A few more experiences like the escalation of costs of the Channel Tunnel (+72%), of the Jubilee Line Extension and the Heathrow Express Rail Link may yet influence the financial world to follow the more enlightened examples born from the ashes of disaster and from other quarters. Chapter 8 explores the circumstances and the explanation of such outcomes.

More generally, major tunnel projects administered by an Engineer, charged to manage the *design* process, incorporating the best elements of risk-sharing and of applying foresight to the heading off of incipient problems, will continue to be seen as a superior solution to that of the expensive *macho* project management that remains in favour from the unenlightened circles of finance. It is also probable that a small Dispute Resolution Board (Section 8.5.2) of experienced (practical not academic) engineers (not lawyers), kept well informed of the development of the project throughout the construction phase, will be seen as a reserved reminder of the over-riding virtue of objectivity of professionalism.

Chapter 8

Hazards, Disputes and their Resolution

They're funny things, Accidents. You never have them until you are having them.

The House at Pooh Corner, A.A. Milne.

8.1 Introduction

Defects in design, in materials, in specification of materials and workmanship, in standards of workmanship or in any combination of these features – and it is most frequently that mutual incompatibility or misunderstanding lies at the root of a problem – demand rigorous control. The costs of rebuilding or in finding alternative solutions in tunnelling are likely to be high, particularly on account of the linear process of construction and constraints on access to working faces. There is no better safeguard than a continuity in direction of the project by an engineer perceptive to the nature of potential hazards and how they may be anticipated. This, however frequently disregarded, is no new concept. Muir Wood (1975b) states:

> The processes of planning, design, construction and maintenance of tunnels are (or should be) closely inter-related. In certain countries, current practices tend towards artificial barriers and discontinuities between certain of these aspects, particularly that of time separation between design and construction; it is not surprising there to find a high frequency of avoidable hazards and frustrated contracts.

Where this lack of continuity exists, or where reliance is placed on oversight by a central administrator in the place of participation by an experienced engineer, readily avoidable and expensive failures to meet objectives will continue to occur at an unacceptable frequency. The lawyers will continue to seek scapegoats without concern or understanding of the fact that a fundamental cause has been an absence of attention to the principles of *design* (Chapter 2).

There is a present tendency in the United Kingdom to regard Construction Design Management (CDM), described in Section 2.4, as in some way compensating for the absence of informed engineering continuity of management. This is to misunderstand the function of CDM. CDM, particularly where it is undertaken through competitive tendering, may be reduced to a review of stated proposals for methods of working, and in provision of a safety plan by the 'Planning Supervisor'. The review may have good intentions but these may take little regard of the manner in which decisions are taken and the response to slight but important departures from routine operations or from stated expectations about the ground. The Planning Supervisor does not provide continuity in supervision – in fact, some never visit the working site. CDM plays no part in ensuring that the objectives of the project, as designed and built, will be achieved. If, for example, a supposedly watertight tunnel leaks, this will only concern CDM if the leakage is life-threatening.

As explained in Chapter 2, hazards may be responsible for physical deficiencies in construction leading to:

- tunnel collapses or distortions;
- flooding;
- fire or explosion;
- injuries or accidents arising from working practices.

External effects include damage to people, the environment, property or services, having one or more contributory causes:

- consequences of tunnel collapse or flooding;
- ground movement causing direct damage (Section 5.3);
- variations to watertable;
- effects of compressed air or chemical agents travelling through the ground.

Non-physical risks concern the failure of the project to meet the objectives of the Parties concerned.

After the event, the immediate circumstances leading to defects are usually capable of being established, often as the result of some degree of specific investigation. If the existence of a particular hazard had been considered earlier, the conclusion will be that the effect could and should have been avoided. It is here that a risk strategy, described in Section 2.1, would have provided a powerful tool in guiding procedures and saving loss that may have been incurred in life, money or time.

There is no attempt here to provide a taxonomy of defects in tunnelling (see, for example, Muir Wood 1975b; Pelizza and Grasso 1998) but rather to consider a few case histories as indicators of the features that may make a project accident-prone and of the positive features that may guide future tunnel project management.

It is frequently supposed that tunnels are subjected to their most demanding conditions during the period of construction. While this is generally true, there remain aspects of longer-term structural security which may only become evident during the passage of time, or in relation to specific features of operation or maintenance. Examples are included in Section 8.4 below.

The important issue overall is once again to emphasise the attitude of mind of all those concerned in directing the operation of construction, who should, in the words of Sir Harold Harding, be 'always prepared to be surprised but not astonished'. The same author also advises (Harding, 1981): 'The moral is that there is no simple answer to tunnelling, but that it is a matter of infinite variety undreamed of by the layman and the administrator.' The essential attribute of the tunnelling engineer, whether or not formal Observational Techniques are adopted (Section 2.6), is that of informed observation which is often the key to forestalling risks.

Defects in tunnelling are too often explained by their most immediate or proximate causes, e.g. the immediate reason for a serious rock-fall. It is instructive to pursue the fundamental causes, which may relate to unclear allocation of responsibility, to lack of communication or to absence of appreciation of the importance of providing specific information to those concerned. A number of minor problems in tunnelling, resolved during the course of construction, but carrying pointers as to how to avoid similar recurrences, are described elsewhere in this book adjacent to discussion of the practices concerned.

8.2 Hazards in construction

Many of the hazards in construction arise from:

* lack of relevant information about the ground;
* failure to express features of geology in aspects or terms significant or comprehensible for engineering;
* misunderstanding of the information available or of its import, occasionally aided by the inadvertent concealment of relevant information from those concerned with designing or managing the scheme of construction.

The safety of the scheme of construction depends on the validity of the ground model (Appendix 5F). Thus the purpose of the ground model should be not only to provide the basis of a generally satisfactory scheme of construction but also to indicate the nature of potential hazards for which special precautions may be needed. In a rock tunnel, for example, the variability of the rock may point to the local presence of loose rock needing to be identified in advance so that local measures may be taken to avoid risk of rock falls. In the more serious circumstances where a fall might trigger progressive failure along a tunnel, there may be the need for periodical local lengths of greater robustness of support, each associated with a formal break to limit spread of collapse. This type of failure becomes more prevalent as the Sprayed Concrete Lined (SCL) tunnel has its use extended to a wider range of types of ground, since a buckling or internal bending failure is most likely to occur with a thin lining of variable geometry, where a high radius of curvature may coincide with poor workmanship and high uneven external ground load to provide the initial failure mechanism (but see Chapter 9 for another example). Emphasis needs constantly to be given to understanding the mechanisms of potential collapse rather than placing reliance on rules of a doctrinal nature, which may be related to a particular form of construction or to a numerical ground quality designation, but which provide little or no guidance on variability or the nature of associated local risks of collapse.

Where probing ahead of the face is undertaken for a rock tunnel, any looseness of rock that may be encountered will be associated with open jointing and hence, below the water-table, with increase of inflow of water. Probing should have specific objectives, with an understanding of the features being sought and for any appropriate precautions. For example, if localised zones of water under high pressure are to be located, collaring of the probe hole will be necessary, equipped with gate valves to staunch inflow. Where running ground may be encountered, the needs for control and for consequential effects of uncontrolled inflow should have been anticipated. In strong rock, where TBMs are used, probe holes should be located outside the area of excavation to avoid obstruction of cutters by abandoned drill rods or bits.

Where reliance is placed on ground treatment to provide stable ground, tests should be conducted to ensure that the objectives have been achieved before opening up the ground. A case described by Muir Wood (1975b) is that of a shaft to be sunk by underpinning through a depth of gravel above clay into which the shaft was to be sealed. The gravel was treated by silicate injection. The bottom subsequently 'blew' and the shaft flooded, requiring an enveloping sheet-piled box for its recovery. The cause of the problem was later identified as a layer of sand at the base of the gravels too fine to be penetrated by the injection. A trial hole could have located the problem and allowed a safe method of working to be devised. If the problem had been suspected from the outset, by detailed investigation, an alternative form of construction could have been adopted initially, for example by sinking the shaft as an open caisson, with water pumped out only after achieving adequate penetration by the cutting edge below the aquifer.

Several instances of problems have occurred in breaking out from a shaft at the beginning of a tunnel drive. The general problem, experienced in several different varieties, arises from instability of ground associated with its disturbance during shaft sinking, coupled with inadequacies of support in the eye through which the tunnel is to be constructed through all stages of the work. One such example was experienced in Mexico City (Muir Wood 1975b) in highly thixotropic clay, sensitive in consequence to disturbance.

Elementary design faults in tunnel arch support include the failure of foot-blocks on account of the limited strength of the ground and of neglect of weakening caused by uncoordinated local excavation, e.g. a drainage channel. Where arches are to be supported as the bench to a top heading is removed, leg extensions must be in place and loaded prior to undermining the local section of bench support. The lateral stability of steel arches always needs to be considered in relation to the degree of blocking against the ground. Where an arch has straight legs, neither the rock nor the arch in this vicinity benefit from arching and need analysis of the possibility of ground instability and of distortion of the arch.

Small inflows of water may, in particularly sensitive circumstances, have consequences disproportionate to the quantities concerned. Examples of this nature have occurred in Switzerland, as described in Section 5.3 affecting an arch dam, and in Oslo as a result of a small degree of underdrainage of sensitive clay by tunnelling in the underlying rock, necessitating strict control of leakage to within 5 litres/minute/100 metres length of tunnel (Jøsang 1980). Underdrainage of normally consolidated sediments in Hong Kong as a result of rock tunnelling (Munro 1997) is also described in Section 5.3.

When problems threaten, the prospect of a simple rational solution may depend critically on the contractual terms and relationships between the Parties. Thus, where the expectations of all Parties appeared to be for a dry tunnel, in fact it was so wet, with exposure of plant to salt water, that major modifications would be needed to the TBMs in use to adapt to the conditions. Moreover, construction time would then grossly exceed the specified period. All geological risk had been placed with the Contractor who saw no prospect of renegotiation on the grounds of unforeseen conditions. With so much at issue, with no prospect of agreement, the cost, diversion of effort and delay as a result of litigation will result in far greater overall cost to all Parties, including the Employer, than would have resulted from a more equitable allocation of risk in the first place.

There is a common phrase – 'technical before contractual' – which implies that solution to the engineering problem, in a tunnel or elsewhere, will normally not wait for debate on the minutiae of liability. Such an approach assumes, however, an equitable basis for subsequent compensation and a cooperative relationship between those concerned in assessment. There may be no such basis, as in the circumstance of excessive risk being imposed on the Contractor and where information provided by the Employer may have been misleading or incomplete. Almost as bad is the situation where the Engineer, in today's litigious climate, is conscious that a ruling in the favour of the Contractor may lead to the Engineer being pursued in the court by the Employer, his Client, for an aspect of alleged negligence which may have contributed to the problem. Once again, this usually arises on account of an Engineer accepting responsibilities for tunnelling beyond his grasp.

A hydro-electric project on a Pacific island was designed to be constructed in hard volcanic rocks overlain by sediments with known presence of basalt dykes and sills. Unexpected problems were encountered in raising an inclined shaft, attributed by an Austrian professor serving on a Consultants Board to excessive horizontal stresses. The shaft was being raised by a diesel-driven shaft raiser. On investigation, it became evident that destruction of ventilation trunking as a consequence of rock falls in the shaft created so unpleasant an atmosphere that no serious geological examination had been undertaken of the shaft. A more intrepid engineering geologist rode on the roof of the Alimak shaft-raiser obtaining video film which revealed that the geological structure had been essentially misinterpreted. In fact, the shaft was virtually following a steeply inclined basaltic flow transverse to the expected direction of such an intrusion, associated with weathering and heavy fracture. This unsuspected feature was the cause of much avoidable cost. The obvious solution, suggested by the Contractor, prior to knowledge of the cause of the problem, would have been to substitute a vertical shaft – and here it may be worth commenting that inclined shafts introduce a number of complications in construction sensitive to uncertain geological factors. This proposal had been rejected by the Engineer, on the grounds that the change would inevitably entail increased costs, on account of head losses, in operation. In fact, this issue could readily have been compensated by local increase in tunnel and shaft areas, with much saving in construction costs and time. This is an example of addressing the wrong problem and failing to perceive the right solution, both features obvious in hindsight, the latter also in foresight.

On 21 October 1994 a major collapse of station tunnels under construction within Heathrow Airport for the Heathrow Express Rail Link, using SCL techniques, caused considerable surface subsidence and damage, but fortunately no personal injury. The incident is described in Chapter 9. Only subsequent to this collapse was a management plan introduced which attended to the needs for continuity and coordination throughout the processes of design and construction. The Jubilee Line Extension, under construction at the time and using similar tunnelling techniques, made comparable changes in project management without which it is possible that comparable experiences could have occurred, in circumstances more susceptible to risking life.

While compressed air tunnelling has declined in recent years in response to medical concerns, examples of 'blows' continue to occur, where the air finds a path of escape to the surface (Muir Wood 1975b). The Clyde Tunnel (Morgan *et al.* 1965) was constructed in compressed air, the crown of the tunnel passing beneath a timber jetty used over many years as a ferry berth (Figure 8.1). As-built drawings by its nineteenth century contractor indicated the lengths of driven timber piles. The Author made the excusable error of supposing that the contractor would not have under-estimated the length of pile (on which payment would have been based), but this in fact was the case. The tunnel passed in consequence too close to the shoe of the pile. Constant buffeting over the years by berthing ferries had tended to loosen the piles which provided a preferential path for escape of compressed air, the path enlarged by the mechanism of upward flow of air. The resulting loss of ground into the tunnel was contained without injury, the cavity filled, ground treatment undertaken and tunnelling resumed. The containment of compressed air depends on the degree of homogeneity and continuity of the ground; any disturbance to this pattern needs to be treated with suspicion and investigated. The compressed air should be prevented from accumulating at any level at a pressure equal or greater than the overburden. Where compressed air is to be used, the stratigraphical sequence should be studied in order to consider the possibility of the air leaking from the tunnel being trapped beneath a relatively impermeable layer. It is more surprising to find examples of compressed air failures whose primary cause has been that of air pressure inside the tunnel exceeding that of the overburden.

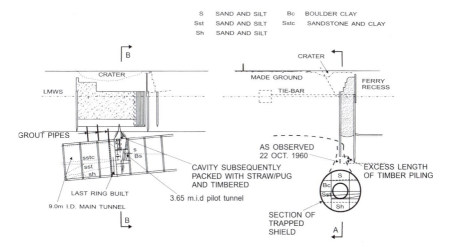

Figure 8.1 Clyde Tunnel blow in West Tunnel on 22 October 1960.

Another feature associated with compressed air arises where the air is driven through organic or chemically reducing soils causing deoxygenation, from a face under compressed air to another in free air or in compressed air at a lower pressure. A more bizarre example of a comparable phenomenon was that of the driving of the first RER Metro line in Paris from Etoile towards the Pont de Neuilly. A shield was used with compressed air confined to a compartment in the face of the shield. High temperatures called for investigation which disclosed that pyrites nodules in the ground were being oxidised by escaping air, leading to production of sulphuric acid, in its turn causing accelerated corrosion of the plating of the shield (Kerisel 1969). Other comparable incidents have been reported elsewhere in natural and contaminated ground.

Many of the chemical resins used as special expedients for ground treatment of fine soils or for sealing fissures of low aperture in rock are highly toxic and require to be handled with appropriate care. Such precautions are widely understood and the less readily controlled problem occurs when they migrate to another cavity or to the surface as described in Section 6.4.

Many incidents of greater or lesser significance have resulted from a tunnel encountering open boreholes, wells and geological features filled with water or loose water-bearing soil. Where past ownership of land is involved, it may be possible by consulting old maps and documents to determine the likely location of wells. There should be available records of recent boreholes. For the Channel Tunnel, while boreholes sunk for the tunnel itself were deliberately offset from the tunnel line (but this was periodically changed) and had generally been conscientiously back-filled with a bentonite-cement mixture to avoid risk of flooding the tunnel, the positions of other boreholes sunk for hydrocarbon prospection had, so far as was known, been left open and constituted an unknown hazard to the Tunnel. Fortunately these were situated predominantly on the French side of the Channel, where the TBMs were designed to operate in closed or open-face mode. None was encountered during tunnelling. This is a hazard which may merit consideration elsewhere in circumstances where exploration for other purposes, lacking a discipline of effective back-filling of boreholes, may have preceded a tunnel project.

So-called 'anomalies' encountered in London clay of Eocene age have been attributed by different authors to different origins. The upper part of each feature has the form of a non-circular funnel or basin with major and minor axes variable around mean values of 50–100 m. Below, a more irregular shape has been found, partially steep-sided, which may penetrate the base of the clay, where underlying strata indicate some degree of resulting uplift. An account by Hutchinson (1991) records 31 examples, referring to drift-filled hollows in the surface of the London clay between 90 m and 470 m across and up to 30 m deep as scour hollows

near the confluence of tributaries to a main river course at the time of
their formation. Hutchinson also concludes that, where the anomalies have
penetrated the London clay, these features are usually found near the
tapering edge of the clay (< 35 m thickness) and where artesian water
would have been freeflowing, the lower strata thrust upwards as a result
of excess of hydraulic pressure from beneath. Hutchinson provides a
taxonomy of five different forms of anomaly associated with the London
clay, the formation of the processes mostly associated with cold climatic
periods or inter-glacial periods. A 6.1 km long, 2.5 m diameter tunnel
between Wraysbury and Iver in the Thames Valley encountered one such
anomaly in December 1983 at a depth of about 30 m. After consideration of alternatives, ground treatment through a pattern of boreholes from
the surface, of the unstable mixture of sand, silt and clay was attempted,
combining *claquage* grouting using bentonite-cement with permeation
grouting with a silicate grout with ester hardener. This was unsuccessful,
indicated by failure of the silicate grout to penetrate some of the
water-bearing ground and by partial absence of setting of the grout. In
consequence, freezing by liquid nitrogen was undertaken through a series
of vertical holes on a grid of 1 m × 1.2 m, based on a 'rolling freeze',
freezing proceeding with the advance of the tunnel. Periodical local problems were caused by irregular inclination of some of the freeze holes.

The confined space within a tunnel is itself a contributory hazard. Where
the completed size of a small-diameter (say < 3 m) tunnel is specified,
consideration should be given to the overall benefit of some increase for
safety, efficiency and economy, particularly for a TBM tunnel, possibly
requiring provisions for special expedients and for ground support. There
should for example be space for a continuous dry walkway for man access
to the face, alongside any rail track. Accidents have resulted, for example,
from a locomotive striking a bottle of propane gas stored too close to
the track in a confined tunnel. The provisions for flow of ingress of water
should be related to tunnel gradient. Where a TBM is used, regard must
also be given to the safety of reasonable access for those who may be
required to undertake maintenance or ground exploration ahead of the
face of the machine. Care needs to be taken in the routeing of tunnel
services during construction. Otherwise a single incident of, for example,
a derailed skip, may cut off power supplies with serious consequences.
The design of mechanical plant should not entail risk of injury or damage
as a result of a power failure, e.g. by the sudden drop of a load. These
are examples of issues to be considered, alongside Codes of Practice and
statutory health and safety regulations for work in tunnels to be undertaken to exemplary standards and probably overall economy.

A number of other examples of tunnelling defects are described elsewhere
in this book. The nature of possible emergencies needs constantly to be
reviewed and the procedures rehearsed. For example, smoke associated

with a fire is a major hazard to life and an obstruction to efficient fire-fighting as a consequence of poor visibility.

Fire precautions in tunnels during construction need to take account of specific features contributing to the hazard such as HT electrical circuits and flammable hydraulic fluids (Tait and Høj 1996). The resort to intrinsically safe electrical equipment and non-flammable fluids imposes high costs, greater requirements for space and needs strong reason for justification, against the alternative of rigorous fire safety measures. The effects of fire on the permanent structure is described in Section 8.4. Working in compressed air adds to the risk of fire in two respects: first by the increased availability of oxygen; second, by any incipient smouldering, of timber support for example, to be kindled by the draught of escaping air.

8.3 Methane

The danger of methane in tunnelling calls for special mention. As methane is an odourless, invisible gas, circumstances leading to explosion may develop without preliminary warning where steps have not been taken to identify the potential conditions for methane to be found. Furthermore, the solubility of methane in water may allow invasion of methane into a space apparently isolated from gas entry from the ground. This gas may then come out of solution. Methane may have a direct biological source from the breakdown of organic matter or it may be of geological, i.e. palaeobiological, origin. Methane causes an explosive mixture in air at concentrations (by weight) between the Lower Explosive Limit (LEL) of ~5% and Upper Explosive Limit (UEL) of ~15%.

The methane explosion at Abbeystead in 1984 (Health and Safety Executive 1985) illustrates a number of features of methane risk, providing a cautionary tale for more generalised application. The tunnel (Figure 8.2), for conveyance of water from the River Lune to the River Wyre in North-West England passes through mudstone, siltstone and sandstone rocks of Namurian (Lower Carboniferous) age. The possible inference of the presence of hydrocarbons in the vicinity of the tunnel as a consequence of a 'commercially confidential' borehole at Whitmoor (7 km away from the tunnel) had not been considered at the time of planning the project. As a result, only routine detection of methane had been undertaken during construction, using portable instruments unlikely to detect the presence of methane in a well-ventilated tunnel, but the question remains arguable as to whether some aspect of construction, involving for example the local lowering of a perched water-table, might have been a trigger to start seepage of methane, in which case the tunnel would have been virtually free of methane during early stages of construction. Environmental considerations led to the air vent from the tunnel ventilation valves being kept below ground surface, discharging into the wet-room of the valve house (Figure 8.3).

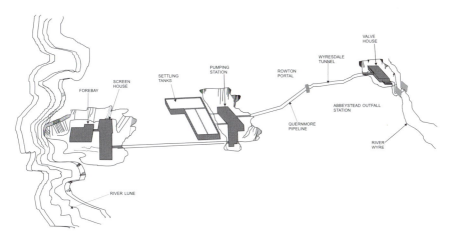

Figure 8.2 Lune–Wyre transfer scheme.

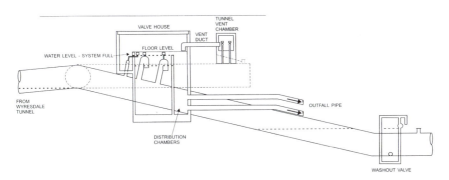

Figure 8.3 Schematic arrangement of outfall works at Abbeystead (after Health and Safety Executive 1985).

An explosion in May 1984 in the valve house, near to the point of discharge into the River Wyre at Abbeystead (Orr *et al.* 1991), at the time of an organised visit by the public, caused loss of life and major structural damage. Investigation disclosed an operating practice, when no pumping had occurred over extended periods, of cracking open a wash-out valve to avoid development of stagnant water in a cul-de-sac portion of the tunnel. There was a certain amount of nett income of ground-water to the tunnel but when the outflow through the wash-out valve exceeded the inflow, this action had the effect of lowering the water level in the upper part of the tunnel system, forming an air void in consequence. When pumping resumed, the air expelled from this void into the wet-room contained a high concentration of methane which, on this occasion, created an explosive mixture.

A detailed study was made of the methane transmission paths, leading to these general conclusions illustrated diagrammatically by Figure 8.4:

1. The methane is derived from source rocks at considerable depth.
2. The methane migrates upwards, along joints in the rock, driven by a 'gas-lift pump' mechanism (reducing pressure causing the methane concentration to exceed the equilibrium saturation level in the water and to come out of solution, thus reducing the overall density of the combined water and gas and hence the mechanism for the upward lift) to a trap beneath a mudrock anticline in the vicinity of the tunnel.
3. Methane enters the tunnel partially as a gas (~50%?), partially in solution, at a mean rate of about 8 kg/day, the rate varying inversely with barometric pressure and with other factors only partially understood. The total rate of inflow of methane was based on analysis of gas flow and water flow over an extended period but the proportion entering in solution is inferred from direct measurement when the tunnel was dewatered.

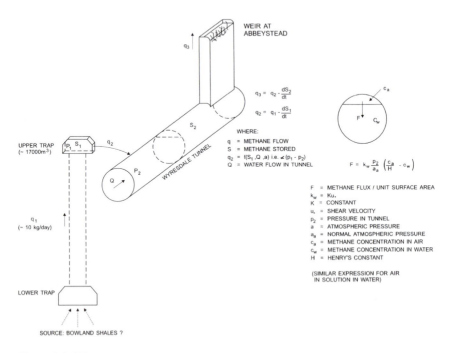

Figure 8.4 Abbeystead: postulated mechanism of methane flow.

An account of the investigation following the explosion is described by Orr *et al.* (1991), partly based on geophysical prospection, partly on evidence from the approximately exponential rate of reduction in methane entry into the tunnel immediately following dewatering, from an initial maximum in excess of 150 kg/day.

Since the factors controlling absorption and release of methane between water and air are not readily accessible to tunnelling engineers, a brief account is set out below. Except in perfectly still water, where molecular diffusion would control, the flux rate F (i.e. rate of transfer of methane per unit area of the water surface into or out of solution) is controlled by turbulent diffusion and, for a gas such as methane which does not ionise in water, is given by:

$$F = K_w(c_w - c_a/H)$$

where: $K_w = Ku$, where K is a constant and $u = (\tau/\rho)^{1/2}$ is shear velocity, c_w = concentration of gas dissolved in water, c_a = concentration of gas in air, H = Henry's constant, τ = shear stress, and ρ = density of water.

At NTP, for 1 bar pressure of methane, i.e. 100% in air, for equilibrium c_w = 27.4 mg/litre, with the value of c_w increasing as temperature reduces and as pressure increases. For flow in a tunnel, τ is calculated from friction at the tunnel wall for rate of the flow of water. For a particular rate of flow therefore, where the concentration of methane in air is small enough to be neglected, $F \propto c_w$. The flux rate F will increase with increasing rate of flow, where other factors remain constant.

Water diffusing from a water surface may give rise to a methane level in air of 5%, the LEL for a concentration in the water of $5/100 \times 27.4$ mg/litre, i.e. ~1.37 mg/litre (adjusted to prevailing temperature and pressure). During intermission of several weeks between periods of transfer, the body of stagnant water in the Lune–Wyre pipeline and tunnel absorbed considerable quantities of methane, much of which came out of solution as the water cascaded over a weir in the wet-room of the valve-house at Abbeystead. Therefore, even with redesign of the air venting arrangements, the wet-room has, as an additional precaution, been opened up to the air.

As a footnote to Abbeystead, it is instructive to summarise the legal aspect. A charge of negligence was tried in the court, with considerable technical evidence on the factors which contributed, without prior appreciation of risk by the Parties concerned, to the explosion. The trial judge, who indicated a wide measure of understanding the technical issues, nevertheless found Employer, Engineer and Contractor all guilty of negligence. Under English law, compensation to those affected depended on proof of negligence. What was a little surprising was that the Contractor should have been found to have failed to test for methane to an acceptable standard; tunnelling practice at that time would not have required such testing to be undertaken unless there were reason to suspect the presence of methane and, as stated above, the chance of detection in a well-ventilated tunnel by a standard instrument would be remote. The Court of Appeal dismissed the charge against the Contractor and considered that the Employer was entitled to rely upon care in design and supervision of the project by the Engineer. Two judges of the Court of Appeal found the Engineer guilty of negligence but Lord Justice Bingham dissented by reference to the test of professional negligence, based on such precedents as the statement by the judge in the often cited case of Bolam v Friern Hospital Management Committee of 1957 which contains this phrase:

'A man need not possess the highest expert skill at the risk of being found negligent. It is well established law that it is sufficient if he exercises the ordinary skill of an ordinary competent man exercising that particular art'.

For an engineer, comparison is to be made with the 'ordinary competent man' (or woman) of the class to which the particular engineer claims to belong. The question needs therefore to be asked in relation to alleged negligence as to the special skills and the standard of performance of such skills on which appointment to the project would be based in order to establish this 'class'.

It is probable that many tunnels with methane entry rates comparable to those of Abbeystead are recorded as having been free of methane. The portable instruments that have customarily been used to check for methane during tunnel construction have a sensitivity that permits concentrations of no less than about 10% of the LEL to be recorded, i.e. 0.5%. Such a concentration would only be recorded in a well-ventilated tunnel with a massive inflow or in a particularly stagnant area of the tunnel. There is then no explicit assurance that, during periodical stoppages or in the event of power failure, local dangerous concentrations of methane will not occur. A surer means of safeguard, where potential danger from methane cannot be dismissed, is by precise measurements of methane concentration in the air discharged by the tunnel construction ventilation system; knowledge of the rate of flow will then allow rate of methane inflow to be calculated. As the tunnel advances, such measurements should be repeated if the risk of migration from a methane source cannot be disregarded. This situation occurred during the construction of an outfall tunnel in Sydney in permo-triassic rocks overlying coal measures. When precise measurements were made, in a tunnel hitherto believed to be free of methane, a rate of entry of methane at a rate similar to that of Abbeystead was established. Provided the potential risks are understood, control of minor inflow of methane should follow coal mining practice by reliance on assured dilution. The remarkable explosion in the diversion tunnel at Furnas, Brazil (Lyra and MacGregor 1967) was caused by methane derived from rotting vegetation, a more readily controlled source, once the potential hazard was identified.

During the construction of Brunel's Thames Tunnel (Muir Wood 1994a) considerable quantities of methane and of hydrogen sulphide entered the tunnel with sewage-contaminated ground-water. The methane caused periodical flares from the tunnel face, but the widespread use of candles to light the works possibly contributed, by igniting the gas at the point of entry, to the absence of the local gas concentrations coming within the explosive limits (~5–15%).

Where tunnels traverse, or are close to, landfill sites, the potential risk of methane needs to be anticipated. A cut-and-cover tunnel (as for the Heathrow Express, Chapter 9) may be provided with an appropriate protection by a thick clay cover with external venting and drainage or, where greater potential danger threatens, by a more elaborate system incorporating polymeric sheeting.

Where any inflow of water occurs, through the tunnel lining or through a portal, the possibility of methane-bearing water must be foreseen, either requiring continuous dilution, control at source or, where internal ventilation is adequate, design of gullies or interceptors to ensure that their local ventilation is sufficient to ensure no danger from methane accumulating above the water surface as it comes out of solution. Comparable precautions are needed in tunnels in which there may, in operation, be spillages of inflammable or toxic liquids, including the provision of periodical U-traps to contain spread of flame.

8.4 Defects during operation

Certain of the hazards described in Sections 8.2 and 8.3 may only become apparent after the elapse of time such that the tunnel is already in service. The consequences will be correspondingly more serious. There are also many examples of natural deterioration with time, at a rate in excess of that expected, on account of undetected factors such as particularly aggressive ground-water. Such latent problems, involving as they do major direct and indirect costs, should be foreseen as possibilities and systematically eliminated as the result of specific investigation at the time of initial planning or taken into account in estimating the life costs of a particular option for a project. Many problems have been experienced of corrosion of reinforced concrete tunnel linings by saline water, a problem touched upon in Chapter 5.

A number of brick and masonry railway tunnels were constructed in the nineteenth century in stiff clay or marl which did not appear to the engineers of the time to require invert arches. The passage of time, accompanying reduced confining stress aided by repeated cycles of stress reversal from rail traffic, has weakened the clay beneath the tracks and in consequence impaired the stability of the footings of the arch. Remedial work may entail the provision of a new invert in short lengths accompanied by temporary propping between the footings. The Bopeep Tunnel (Campion 1967) provides an example of a tunnel collapse when such remedial work was not undertaken in time, leading to a parallel-sided block failure extending to the ground surface.

Where electrification or other form of upgrading has required the lowering of railway track, resistance against inward movement of the footings of an arch tunnel constructed without invert is reduced and collapse may occur. An instance of this nature affected a short, shallow brick tunnel in Glasgow, where water services had been built into the tunnel arch below the carriageway of a road crossing the railway. Here the process of collapse was accelerated by fracture of the water-main. Legal argument, of little real concern to the public, then ensued for some months between the two public bodies concerned as to which was responsible. This was likely to be inconclusive since the two events, tunnel movement and pipe fracture, were mutually provocative in a progressive manner, incipient movement coincident with increasing leakage.

An instance in Australia revealed, only at the time of proposed electrification, when clearances became vital, distortion of a brick tunnel in Geelong. Examination of the history of construction of the original tunnel in 1875 revealed that the distorted sections coincided with the positions of intermediate working shafts. Almost certainly, squatting of the tunnel had occurred as a result of inadequate compaction of backfilling in the working spaces formed at the foot of each shaft, with associated dropping of the crown of the tunnel and settlement of fill in the shaft. In all probability this had occurred over many years, being only noticeable as the result of an accurate survey.

Egger (1996) describes typical problems of swelling gypsum affecting tunnelling in Stuttgart. Intact anhydrite ($CaSO_4$) at depth shows little tendency for swelling but towards the surface of the ground or where for other reason the anhydrite is fractured, access of water permits hydration to gypsum ($CaSO_4.2H_2O$) with associated swelling. Swelling potential of gypsum has been measured between 1 and 4 MPa, which if unresisted may result in an increase of volume of about 60%. The most common form of swelling damage by gypsum has caused invert heave. Where the threat exists, countermeasures have been either to seal off the possible access by water (usually found to be a difficult operation) or, more positively, to construct a massive invert capable of resisting swelling. Gysel (1977) undertook an early analysis of the approach to tunnel design in such circumstances.

A failure of a single-track rail tunnel dating from 1827 occurred beneath the University of Kent at Canterbury in 1974. The tunnel had a cross section of about 4 m × 6 m with the major axis vertical. It is interesting that M.I. Brunel commented to his son, I.K. Brunel (Harding 1981) in 1836 on a tunnel of identical profile, with vertical walls between semi-circular arch and invert: 'They have not, I conceive, given sufficient arching to the sides ...' The shape was doubtless prompted from simplicity in centring by comparison with a preferable ellipse. At the area of failure over a length of about 30 m, the tunnel was in London clay and passed beneath the Cornwallis Building of the University. The line had been closed to traffic in 1952 and from 1964 the ends of the tunnel closed by bulkheads. There had been evidence from inspections since 1952 of bulging of the sidewalls but the precise mechanism of incipient failure was not identified. While the tunnel remained ventilated, evaporation from the surface of the brickwork would have enhanced the strength of the brick-work and would have applied a suction to the clay around the tunnel, thereby increasing its effective strength. The clay immediately around the tunnel was found to have its moisture content increased by about 10% and a wedge-type failure is surmised as the initiating mechanism (Vaughan et al. 1983). Increasing inward bulging of the brickwork would cause increasingly concentrated loading on weakened brickwork. At first, in 1974, two breaches were observed in the lining, which later joined together and caused subsidence of a vertical-sided block of clay.

A spectacular fire occurred on 18 November 1996 in the Channel Tunnel, (Brux 1997) notwithstanding the elaborate safety precautions built into the operation of this link. A lorry (HGV) loaded with expanded polystyrene was ignited outside the tunnel, possibly by a flaming missile thrown by a striker. The policy for dealing with a fire detected on a train was for the transit of the tunnel to be completed and the fire extinguished outside the tunnel. On this occasion, the fire affected a warning circuit operated by a limit-switch relating to the vehicle-loading mechanism. If this were out of its correct stowage position, it could foul the loading gauge appropriate to the tunnel. The train stopped in consequence and the temperature of the fire rapidly caused local melting of the soft metal fixing the power cable to its support. The power cable dropped and the resulting short-circuit cut off power. The train crew and HGV drivers escaped to the service tunnel provided for such eventuality, some affected by smoke since instructions for using the ventilation system to ease escape were slow to take effect. A major contributory cause of this incident was that, initially, the intention had been for the HGV wagons to be enclosed and for any fire to be extin-guished by the automatic release of Halon gas. During wagon design, increasing complexity introduced weight and restrictive interpretation of maximum axle-loading caused the canopy of the wagon to be omitted thereby eliminating such a possibility for fire control; the overall risk does not appear to have been fundamentally reconsidered at this late point in construction of the project. Smoke from the vehicle had been seen by a security guard before the train entered the tunnel but there was no procedure for immediate notification of the train control centre. As a consequence of this incident, several operational changes were made, including provisions more readily to tackle such a fire within the tunnel.

A feature of this fire was that within the tunnel the temperature rose to 1000°C or more and caused spalling of the reinforced concrete lining over a length of about 500 m and for part of this length back to the rear reinforcement in the 400 mm thick segments. Spalling was principally attributed to high pressures from trapped moisture in a dense concrete. Fortunately this incident occurred where no immediate risk of irruption by sea (which could have helped to extinguish the blaze?) or ground was threatened. Repair made use of sprayed concrete. An internal layer of foamed concrete has been recommended as a protection against such damage to a vulnerable tunnel elsewhere.

8.5 Disputes

Disputes are expensive and time-consuming. A tunnel project will entail some degree of uncertainty. The first elementary step, often ignored, particularly where project responsibilities are fragmented, is to identify the causes for uncertainty and to establish that procurement and contractual arrangements take these adequately into account (Chapter 7).

8.5.1 Causes of disputes

The commonest causes for disputes in tunnelling occur where the project has changed, or has appeared to change, in such a way that the rates or terms for undertaking the work are alleged as being no longer applicable. The Contract may or may not make provision for a stated degree of variation. For tunnelling contracts, the most likely single cause for dispute arises from claims for unforeseeable conditions of the ground, affecting the costs of construction and possibly of consequential remedial work. Uncertainty as to the base from which any important change is to be measured is a significant cause for dispute, for immeasurable argument and for rich pickings for the lawyers. The use of Reference Conditions (CIRIA 1978), described in Section 2.6, is a valuable control, used with discretion and adequately related to the features affecting the choice of a specific scheme of construction. It is interesting to find that this same concept has been reinvented in the United States 19 years later under the title *Geotechnical Baseline Report* (ASCE 1997). Where geological interpretation has failed to appreciate the sensitivity of a particular means of construction, a problem may arise when the Contractor elects to adopt such a means. Similar problems may arise where the particular circumstances associated with ground-water have not been understood, or not been revealed by site investigation information available at the time of Tender.

Problems relating to the ground are exacerbated where all geological risk is thrust onto the Contractor. Either exposure by the Contractor to excessive financial risk may result or the cost of the project to the Owner will be very high, or usually both. Since there is then no available evidence as to the base line of expectations, no opportunity for handling unexpected risk by contributions from revisions to planning or redesign of the works, no assurance of compatibility with the selected scheme of construction, uncertainty is multiplied and all Parties suffer. It deserves emphasis that, by imposing all responsibility on the Contractor, the possibility of changes to the project, to help in solving an unexpected problem, will entail a Variation which may undermine the basis of the Contract. If the Contract contains any feature that has deliberately or inadvertently misled the Contractor into supposing a lesser risk, or where relevant information known to the Employer or Engineer has not been made available to the tenderers, the grounds for dispute are established. It is also highly improbable that an adequate hazard assessment will have been undertaken in the circumstances of the Contractor bearing all geological risk. These are all features of poor and expensive project management.

Where the Contract is administered by an Engineer who has been responsible for the planning and design of the project, with the associated studies, he will, on the approach of an incipient problem, be in a position to explore how revisions to layout or concepts of design might assist towards a solution at least cost to the Employer and making best use of the resources of the Contractor. Where the Engineer (or the Supervisor under the New Engineering Contract) has had partial or negligible concern with developing the project, he will be unable readily to propose such a contribution and may not have the powers to do so. An Engineer appointed to a fragmented project, moreover, is likely to find that he, having been appointed under some such title as 'Design Contractor', not enjoying a professional relationship with the Owner, may be the subject of litigation from his own Client, a further barrier to finding a coordinated solution whose antecedent causes lie in the very fragmentation.

A particular category of problems may arise for a Target Contract (Section 7.6). The Target Value should be established on the basis of assumed and agreed Reference Conditions which may then be used for the basis of establishing change. For a simple case, where the basis of estimate is related explicitly to a stated distribution of recognisable objective characteristics of the ground, there should be no problem in revising the value of the Target. If the assumed Reference Conditions are not explicit there may be a problem. A problem also arises for design-and-build projects where it may be suggested that the increase in cost is at least partially caused by the adoption of a scheme of construction intolerant to change in the conditions of the ground. This type of problem may be averted by the undertaking of a hazard assessment prior to acceptance of the scheme of construction.

Where the Contract is controlled by an Engineer, a mechanism is in place to settle problems expeditiously. A serious condition arises when the Engineer fails to behave in an ethically neutral manner. In a notorious case, which did not proceed to arbitration, the Engineer, not foreseeing possible geological hazard, had undertaken to his Client to complete a tunnelling project within a stated budget. When such a hazard intervened, a claim for compensation by the Contractor was rejected.

Problems may arise from the Contract Documents themselves, of which examples are:

1. Incompatibility between different sections of the documents without clear indication of precedence.
2. Inappropriate documentation, caused by inadequate knowledge of tunnelling in the preparation of the documents. A drainage authority, familiar with work near the ground surface, specified forms of grouting of rock at depth applicable to a shallow tunnel in soft ground.
3. Inappropriate assumptions of responsibility by the Engineer for an aspect of construction which is then unsuccessful. There are occasions where the continuity in control between design and construction is essential for an innovative scheme of tunnelling, where it is correct for the experienced Engineer to establish the essential ingredients of the scheme of construction. There are other occasions where inexperience has caused the problems. For the Carsington water supply scheme, where water-bearing shaley mudrocks presented difficult tunnelling conditions, the Engineer assumed responsibility for a vital step in construction, namely the means of assuring a stable tunnel face, without a clear idea as to how this was to be achieved. Specified shields and road-headers obstructed the working faces of each tunnel preventing access to undertake the support and drainage measures which were the appropriate means for control.
4. Vagueness or misleadingly absolute requirements in specification. What does it mean when a tunnel is required to be watertight? Engineers may know what they mean but lawyers may take the requirement literally. CIRIA (1979), for example, provides a set of specific standards of dryness.
5. Impossibility of performance. It may be impossible, without special expedients which are excluded, to comply with stated limits of ground movement. Achievement of the completion period would require a form of construction that would fail to satisfy some other provision of the Contract.

The Channel Tunnel was a hybrid of curious contractual parentage. While the Conditions of Contract between Eurotunnel (Employer) and Trans-Manche-Link (Contractor) were based on FIDIC, variations to this document served to confuse the situation. For example, there was considerable play upon the concept of 'optimisation', generally understood in such a context as establishing the best balance between capital cost and operating cost. The term offers almost infinite scope for interpretation where one Party bears the greater part of the capital costs, the other Party the costs of operation and maintenance, with the absence of provisions for 'transparency' in either respect. The Engineer, essential for the balance of a FIDIC type contract, was transmuted into the somewhat emasculated form of *maître d'oeuvre*, with powers of advice but not of decision, which body was then largely, and unilaterally, absorbed into Eurotunnel. Tunnelling was the subject of a Target Contract, scarcely appropriate when not even the tunnel diameter was known at the time of Tender. Fixed equipment, including Terminal Buildings and equipment throughout, was nominally a Lump Sum, but subject to a great amount of variation as the project developed. Rolling stock was included as a series of Provisional Sums. The Anglo-French Safety Commission, only appointed a year after work had begun, was responsible for setting operational standards, which changed as the project developed. It is remarkable that this project, the subject of so much attention by Governments and by such bodies as the British Major Projects Association, should have demonstrated so many egregious and elementary errors in its contractual constitution. It provides a cautionary tale on an unaffordable scale. The main lesson is that a project of this nature requires a deliberate period for project definition and for examination of the several causes for uncertainty before work is commissioned. No overall loss in time would have resulted for the Channel Tunnel with such an approach. The engineers who contributed so much to the project deserved better from the financial direction of the Project.

8.5.2 Resolution of disputes

Disputes arising from tunnelling projects are liable to be costly, protracted and to absorb much nugatory time and effort of engineers and others better occupied in a productive capacity. The first objective must therefore be to prevent disputes from arising or, where this appears unlikely to be practicable, to provide a short simple system for their resolution.

If we look back to 1978, say, the British contractual system depended on a Contract between Employer and Contractor administered by an independent Engineer who, apart from designing the project, controlled and ruled on questions raised by Contractor (or occasionally the Employer) arising from the terms and conditions of the Contract, for tunnelling often associated with questions of foreseeability of the characteristics of the ground. The best Engineers studied carefully the information available about the ground in order to assist the Contractor in selecting a compatible system of working and to prepare for dealing with aspects of uncertainty which might affect the costs of the preferred scheme of construction. There have always been opportunities to make provisions in the Contract for dealing with important uncertainties, codified, for example, in the use of Reference Conditions (CIRIA 1978) from which to be able to measure departures, to reduce argument over what was reasonably foreseeable. During this period, it was also the prerogative of the Engineer to have access to the basis used by the Contractor in constructing his unit rates so that the same formulae could be used for pricing additional work. It seems that this condition has recently been rediscovered by lawyers under the term 'escrow'. Chapter 7 has discussed the choice of form of procurement appropriate for tunnelling, including the possible modifications to the New Engineering Contract to enable benefit to be derived by tunnelling from its undoubted improvements upon predecessors.

The essence of the operation of this 'traditional' system was the independence of the Engineer. Problems arose when the Engineer, possibly influenced by concern over the source of fees (or salary for an Engineer internal to a Client Authority) or by the possibility that extra costs claimed by the Contractor might be related to deficiencies in design, acted in a defensive manner. Another problem arose where certain large Authorities reduced the powers of the Engineer below those stated in the Contract and thus destroyed the intended equilibrium of the ICE and FIDIC Contracts – a source of special grievance where the Engineer failed to make this explicit to the Contractor. The system depended on mutual trust between the Engineer and his Client. This became sadly eroded when the Client, instead of turning to the Engineer for advice as to how to overcome a problem, would first involve a lawyer, effectively preventing a resolution of the issue at a professional level. The alternative was bound to be expensive and suboptimal. The effects are more profound. All good engineering, introducing innovation to the benefit of the project and its Owner, must entail periodical mistakes, usually insignificant in relation to the project overall. If however each such instance is to provoke intervention of the law, the Engineer and the entire project will tend to be conservative in intention and expensive in out-turn. The consequence has been increased cost, reduced innovation, increased confrontation. This adversary situation has for many years attracted comment in the USA and more recent steps for improvement. As a consequence, it is noticeable that, in the USA, there is a greater chasm between 'design' and 'construction' than in Europe or Japan, leading to greater conservatism and legal intervention. Most innovation in tunnelling has either come from overseas or from plant manufacturers.

The ICE or FIDIC Contract in 1978 included arrangements for Arbitration, in the event that the decision of the Engineer was unacceptable to the Employer or Contractor. The Arbitrator was likely to be an engineer of considerable experience in the type of work, either active or recently retired, capable of judging a situation through the eyes of those entering into such a Contract in the circumstances of the Project concerned.

At around this time changes were already occurring, at first with the Arbitration system. Presentation of the case for each Party was no longer through the responsible engineers but through the intermediary of barristers. As a consequence, the Arbitrator, who had previously relied on external Counsel for advice on legal issues, required a greater degree of legal training and was increasingly drawn from a list of professional Arbitrators remote from the practice of engineering, even if some were engineers *manqués*. The consequences have been greatly protracted and litigious proceedings engaged increasingly with legal issues of marginal bearing of the particular claim and frequently well beyond the limits of thoughts of those who entered into the Contract, lacking the virtues of speed, low cost and a form of rough justice appropriate to the circumstances of civil engineering, particularly of underground, projects. The costs of the process have in consequence frequently been well in excess of the sum at issue and with decreasing benefit of arbitration as opposed to a civil action at law. A statistic for 1997 indicates that only one-sixth of the costs of Arbitration for construction disputes in Britain of £750m, returned to one or other of the Parties to the proceedings. Much of this cost relates to tunnelling and the figures exclude much direct and indirect costs relating to the time of skilled engineers distracted to such a preoccupation.

While the presence of the independent Engineer as administrator of the Contract remains appropriate for projects capable of definition and control within a framework which does not interact unduly with the operating policies of the Employer or with other contracts administered by others, other arrangements discussed in Chapter 7 provide satisfactory alternative systems for the circumstances described. In these arrangements, the Employer takes an active part, even the leading part, in the project administration, through engineers competent to introduce solutions to problems that have significance beyond the Contract, with the overall quest for optimisation uppermost, that is to say optimisation of construction combined with operation. As described in Chapter 7, a flexible approach to the Contract relationships remains essential, in view of inherent uncertainty. Variations of a nature foreseen from the outset can then be introduced to the overall benefit of the project and to achieving the operating desiderata. Ultimately, where 'partnering' occurs (Section 7.6) all such discussions and variations take place within a single team with common incentives for achieving the objectives of the Client's 'business case'. There are many intermediate arrangements, whereby profit derived from a project is related to the success achieved overall. An essential element is that of transparency in all aspects concerning the technical features of the project, provision for interchange on aspects that might be improved and generally the encouragement of professional competence to complement the commercial capability of each participant.

Where there is no independent Engineer (e.g. where the Project has been designed by a team headed by the Employer), or where the powers of the Engineer are curtailed, expediency and economy require the presence of a first point of appeal, able to give a rapid technical decision on matters requiring resolution. There is much evidence to support the benefit of the provision in the Contract for the participation of a Dispute Resolution Board (DRB), or a body of such a function under a similar title. This Board usually comprises three or more engineers of considerable practical project experience (from the author's experience, academics do not make effective DRB members) whose fields of activity are somewhat complementary and relate to work of nature similar to that of the Project. The Board retains a familiarity with the project, by being provided with all contract documents, by periodical visits and meetings with the principal engineers concerned with the project. The DRB is available in consequence to respond rapidly to matters of dispute referred to it by one or other Party. The greatest success of such a Board, dependent on the degree of consideration for equity in the Contract, particularly the provision for risk-sharing, occurs where the mere presence of the DRB encourages the Parties to internal resolution of potential disputes, aware that the DRB will reach conclusions that depend expressly on technical appraisal of the points at issue and not on enlargement of the case into areas of legal intricacy beyond the intentions of the Parties to the Contract. This approach satisfactorily inhibits the legal advisors to the Parties from their incorrigible hankering for forensic complexity.

While the DRB members should be capable of understanding the engineering principles and have a general familiarity with site practices they do not necessarily require to contain between them expertise in every aspect of a possible reference, relying principally upon their ability to understand statements by engineers working within such areas who appear on behalf of the Parties, or in exceptional cases on independent authorities to whom the DRB should have the power to turn. From experience of the system, so long as the proceedings are conducted by engineers, such a need is rare.

Where disputes may develop as a result of differing interpretations between the Parties of features of the Contract, it is highly desirable that reference be made to the DRB before any considerable sum of expenditure has been incurred within the area of contention. Otherwise, liabilities without adequate control may be incurred by one Party, encouraged by legal or other advice that the bill may be passed for settlement by the other Party. The cost has been incurred and there is then no solution other than apportionment of this additional cost, a 'zero-sum' situation (Section 7.2). Moreover, if left to fester, issues accrete and resolution of the original problem becomes increasingly difficult to determine equitably.

While the precise rules of operation of the DRB will depend on the nature of the project, the general pattern may be summarised thus:

1. Members of the DRB should be independent from (or declare to the satisfaction of the Parties) any current relationship with either Party.
2. Membership of the DRB and appointment of Chairman will be agreed between the Parties at the time of appointment of the Contractor.
3. Members of the DRB shall have experience in the type of work of the Project.
4. There should be agreed arrangements for the provision of documents, organisation of periodical meetings and visits to the Works, to familiarise the DRB with the progress of the work and with any incipient problem. It is important to develop a mutual understanding and confidence between the Parties and the DRB.
5. The Parties should make the greatest efforts to settle differences of view, and to define precisely the area which defies resolution, prior to the presentation of a dispute to the DRB.
6. Formal rules will be established for the notification of a Dispute to the DRB, in writing and presentation at a hearing, including responsibility for logistics.
7. The DRB should determine, for agreement with the Parties, the rules of operation in the event of a Dispute including a stated period of, say, 10–14 days after the hearing for the DRB to issue their recommendations to the Parties, with provision for a longer period, by agreement with the Parties, possibly with supplementary hearings, in the event of a complex Dispute.
8. The DRB should strive towards a unanimous decision; where this is impossible, a dissenting view would be provided identifying the specific area of disagreement.
9. There will be a stated period of, say, 10–14 days, for acceptance of the recommendations of the DRB by the Parties, after which the recommendation will be binding on the Parties.
10. It is important that the operation of the DRB and the mechanism for acceptance of its decisions and recommendations should be explicit in the Contract (to avoid complications with insurers and others).

Alternative Dispute Resolution (ADR) has developed and been used in the USA since the early 1980s to overcome the mounting costs of litigation and to provide more relevant means for settling disputes. See for example Thomas *et al.* (1992). ADR takes several forms, which may include DRB, but including also arrangements which introduce a single mediator into discussions between the Parties or a private hearing before a judge. However, ADR generally maintains the expensive and inappropriate feature of a distinct legal flavour. In Britain there is a requirement that a contract shall make explicit provision for dispute resolution.

The merit of DRB in an appropriate risk-sharing form of contract as described above is that it is a practical remedy acceptable to the Parties, but not having meticulous concern with points of law which, while having possible bearing on the Contract from a lawyer's viewpoint, were not influential to the attitudes and behaviour of the Parties. If the DRB is put into a legal setting, its operation is inhibited and the process tends to develop the current failings of Arbitration by complication and diversion into legal byways, of interest to neither Party. Not only will the DRB then become more expensive and long-drawn in its operation, by virtue of its infection by high legal fees and costs, but all aspects will be affected by the legal intermediaries, aided no doubt by another trans-Atlantic import, the 'claims engineer' who comes between the engineers who work on the project and the DRB. The DRB should be seen as a constructive part of a project team with interest in promoting the overall success of the Project in the interest of the Parties to it.

The Channel Tunnel and the Øresund Link, comparable projects based on totally different philosophies of project management, each included provisions for Dispute Resolution.

The dispute arrangements for the Channel Tunnel, established within the legalistic trappings preferred by the dominant banking interests, were dominated by lawyers. Generally remote from the engineers putting their best abilities towards the construction of a demanding complex project, confrontation was the keynote of the references to the Dispute Panel. The Contractor was responsible for the greater part of design of the project, subject to an elaborate system of approval by the Employer. The Dispute Panel was appointed by Eurotunnel (Employer) and Trans-Manche-Link (TML, the Contractor) and was seen as the first stage to Arbitration which, itself, was conducted by lawyers appointed through the International Chamber of Commerce. In such circumstances, issues were only brought to the Panel when large sums of money had already been committed. Any solution was therefore of a 'zero-sum' nature – disregarding the not inconsiderable costs of the large teams assembled for each dispute. This was an unsatisfactory arrangement but at least the Dispute Panel helped to avert the costs, greater possibly by an order of magnitude, that would have been entailed in full-blown Arbitration of each issue.

The Øresund Link, based on a form of contract developed by engineers and including provisions for the sharing of risk, depends for its construction on four main Contracts: the bridge (7.8 km); the immersed tunnel

(3.5 km); dredging and reclamation, including the construction of an artificial island 4 km long, where tunnel and bridge meet through transitions of road and rail. The permanent works were designed by the Owner and his consultants with provisions for variation by the Contractors. Each Contract has a Dispute Review Board whose terms of reference correspond generally to items 1–10 earlier. The Dispute Boards maintain close links with the project, through periodical meetings and visits, principally serving as a reminder of the immediate availability of an objective view on any point that eludes the participants. All discussions are between engineers prompted by common interests, assured by the Contracts, for completion to time and budget of the Project as a whole, as well as the individual Contracts.

This comparison of dispute resolution arrangements represents one difference among many between major projects, dominated by financial interests, who rely on defensive postures protected by what they believe to be fortified positions, and those established by engineers who understand the essential element of cooperation across all those aspects of *design* which should contribute to a well conceived project. Differences in control of time and cost of these two projects should provide the best evidence of the preferred approach. At the present time, whereas the Channel Tunnel was about 72% over budget and about two years late, the Øresund Link at the time of writing appears to be generally within budget and on time. Bankers should understand that a major underground project has negligible value (growing mushrooms?) prior to completion and operating. Instead of insisting on premature timed 'milestones' for payment, conditional on their achievement, which give no protection to finance and may – and do – provoke an over-hasty start, a project, if to be designed by the Contractor, should begin with a period of 18–24 months of deliberate preparation. This could have saved the Channel Tunnel a large proportion of the over-run in cost.

To repeat once again, the essence of the formula for success throughout a tunnelling project depends on the appreciation of the professional demands made upon the participants. Where requirements may be specified with a high degree of certainty, the technical competence of each may be exploited without provision for change in the undertaking. It is the inherent uncertainty in tunnelling which demands the high degree of professionalism, the ability to see beyond the solving of individual technical problems by the pooling of special skills for the benefit of the project overall, the realisation of the Owner's business plan, and a profitable and rewarding experience for all concerned. The lawyer's approach is contrary to this objective, erecting a shell of protective clauses around each participant and leaving each to solve problems in isolation – at the expense of one or more of the other parties. This does not make for good engineering.

Generally, there is a wide degree of available knowledge and experience of how to undertake large underground projects successfully, but there are continuing examples of wilful failure to learn the necessary and sufficient features by those who commission new projects. A main objective of this book is to help to discriminate between good and bad practices in all aspects from the initial concept through to the avoidance of disputes and the resolution of those which are unavoidable.

Chapter 9

Coda: the Heathrow Tunnel collapse

'. . . the curious incident of the dog in the night-time.'
'The dog did nothing in the night-time.'
'That was the curious incident.'
Memoirs of Sherlock Holmes, A. Conan Doyle.

9.1 The context of the project

As this book was about to go to press, the legal action arising from the tunnel collapse at Heathrow Airport in 1994 came to court. In consequence, it is now possible, as a personal view, to recount the circumstances, to identify the contributory factors and to point to lessons to be learned for the future. These lessons are instructive and bear much relevance to the principal themes of other chapters. Following the collapse in October 1994, Britain's Health and Safety Executive (HSE) established an investigating team overseen by a Supervisory Board to which the Author was appointed as an external member (Health and Safety Executive 1996). The Author also subsequently served as Expert Witness to the HSE in the indictment proceedings brought against Contractor and Specialist Consultant under the Health and Safety at Work etc. Act 1974.

The practicability of using systems of tunnelling in London clay based on Informal Support (Chapter 5) had been discussed, generally favourably, for several years. On 17 February 1994, a debate on the subject was organised by the British Tunnelling Society. The Author seconded the motion but insisted that the fragmented system of project management currently in favour by London Underground and the British Airports Authority for the tunnelling projects in London needed to be radically changed. The organisational features for success were then reiterated in September 1994 (Muir Wood 1994b), thus:

'Good tunnelling practice demands continuity and interaction of planning, investigation, conceptual design, detailed design and construction. Each is dependent to a degree on the others. A site investigation, for example, needs to be directed to obtaining information of particular relevance to a specific form of tunnelling; where unexpected features are revealed, the tunnelling strategy may need to be reconsidered and the site investigation appropriately varied. Conceptual design and construction are particularly interdependent since the former may depend upon quite specific features of the latter for success, with the need to ensure that these are rigorously implemented.

Present trends in commissioning tunnelling tend to ignore a condition for good tunnelling: the overall management of the design process. The many engineering activities of a project are subdivided and performed sequentially or separately, with only limited coordination. This ensures that interaction cannot occur and that the specific needs cannot be addressed in the early phases.

The single motive appears to be to ensure fixed costs of each fragmented activity, an objective far removed from obtaining good value for money. The costs may well be fixed – up to a point – but the price for so doing will be high and good tunnelling practice suffers in consequence. The goal of economic tunnelling, which benefits all involved, is effectively prevented.

Moreover there are greatly increased risks of disputes and litigation because of the attempt to unload all responsibilities into construction. This procedure is as good for the legal profession as it is disastrous for good engineering.

Tunnelling methods based on the observational method (ISOM) [i.e. Informal Support based on the Observational Method] are particularly incapable of optimisation where the overall project is fragmented.'

The absence of consideration for the criteria of a system of contract management suitable for tunnelling and dependence on the New Engineering Contract (NEC) (Institution of Civil Engineers 1993) not previously applied to tunnelling, paid inadequate regard to the need for special considerations and provisions to make it suitable for the purpose. In particular, the NEC, reissued in 1995 as the NEC Engineering and Construction Contract (Institution of Civil Engineers 1995), subdivides the traditional function in Britain of the Engineer (Chapter 7) into those of designer, Project Manager and Supervisor coupled with the statement that the functions of the designer are limited 'to develop the design to the point where tenders for construction are to be invited'. The NEC advises that the roles of designer, Project Manager and Supervisor may be combined 'where the objectives of the Employer are served by so doing' but there is no guidance as to the circumstances for such combination. As discussed in relation to the Øresund Link Project in Section 7.1, where responsibility for quality is placed upon the Contractor in a competitive contract, a rigorous assessment is needed to ensure that full provision is made to undertake this role comprehensively and conscientiously, under terms which will be understood by the tenderers. Informal Support is essentially a design-led activity. With the break between pre-Tender design and post-Tender design for these tunnels at Heathrow, there was therefore need to ensure a full understanding of the responsibilities thus accepted by the Contractor and the adequacy of the means to undertake these functions. The evidence that emerged during the hearing indicated that no adequate heed was given to the precautionary measures needed when working in circumstances so unfamiliar to those concerned. It deserves note that such actions would have been required by the 1994 CDM regulations.

Heathrow Express provides a high-speed rail service between London and Heathrow Airport, by means of a spur from the main railway line from Paddington, London to Bristol (Figure 9.1). The spur goes underground in a cut-and-cover tunnel through a landfill site, with precautions against methane entry which included a surround of recompacted clay spoil from the tunnelling, provisions for interception of infiltration, a ventilated filter and capping. The remainder of the running tunnels is constructed predominantly with expanded precast linings, while the Station tunnels at the Central Terminal Area (CTA) and at Terminal 4 make use of Informal Support (NATM) for the Platform and Concourse Tunnels.

Mott MacDonald (MM) were responsible for layout planning and supervision of site investigation. Contracts included responsibility for detailed design and for the option of Informal Support. The Contract was awarded to Balfour Beatty (BB) who appointed Geoconsult (GC) as their specialist NATM consultant. A project management structure was established by BAA, under the name of HEX, including staff seconded from Taylor Woodrow Management Contracting Ltd (TWM) as Project Manager and MM as Lead Designer. This was known as the 'Seamless Team' for project management but it was divorced from the management of construction, all details of which were entrusted to BB with retention by HEX only of powers of periodical 'audit'. MM were responsible for the design of permanent works and as checkers of ground settlement. Separately, MM were appointed by BB to advise on control of settlement by compensation grouting. There were in consequence virtually two separate 'managements' with, between them, only a tenuous link. Potentially – and as events proved in reality – this was hazardous for a novel system of tunnelling in London clay, a system fundamentally dependent on control by design through observation and monitoring, under an unfamiliar form of contract.

In view of the technical novelty, a trial tunnel to the same geometry as the future Platform and Concourse Tunnels had been constructed on the line of a running tunnel at Heathrow, demonstrating three possible forms of construction (Bowers *et al.* 1996):

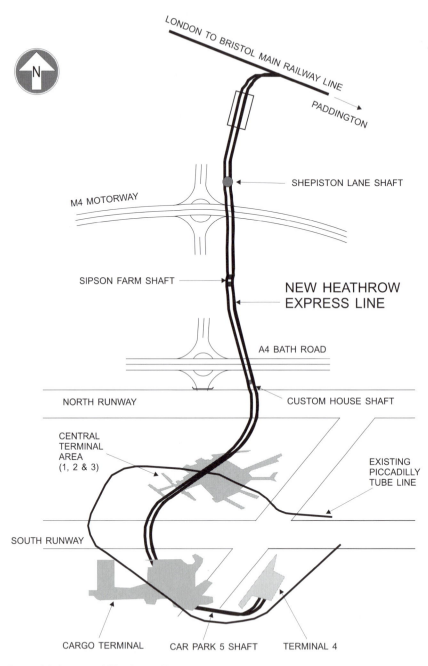

Figure 9.1 Layout of Heathrow Express project.

- Type (i) the use of two side headings with central enlargement.
- Type (ii) the use of a single side heading with enlargement.
- Type (iii) crown heading and bench, with a temporary scroll of shot-crete lining at the base of the crown sidewall ('pretzel' to Austrian engineers, a feature of tunnels without invert in Germany in the early twentieth century).

Each type was constructed for a length of about 30 m. Cover to the tunnel was about 16 m, as opposed to 20 m at CTA. Measurements were taken of lining deflections (convergence), of ground 'stress', of piezometric pressures and ground movements. Immediate surface settlement of about 25 mm was measured for Type (ii), somewhat more for Type (i) and, affected by delayed installation of invert, about 35 mm for Type (iii).

9.2 The project unfolds

BB and GC elected to adopt Type (ii) (see above) for the construction of the Station tunnels at CTA. Work was undertaken by breaking out from the Fuel Depot Shaft previously sunk for access. Initially, in May 1994, work concentrated on the central Concourse Tunnel (Figure 9.2) with the intention of advancing at a rate of 4.2 m per week. For a number of reasons, including material supply for the sprayed concrete (dry process), progress averaging only about 1.5 m per week fell behind programme. BB decided in consequence to start construction of the Down Line Platform Tunnel in September 1994, followed soon thereafter by the Up line Platform Tunnel. As a part of their undertaking BB were liable for making good damage that might be caused by failure to respect settlement limits for structures above the tunnels. These structures included (Figure 9.2) Camborne House and the Piccadilly Line of London Underground.

There is no evidence of a comprehensive risk analysis having been under-taken, addressed specifically to the allocation of responsibilities. This is particularly surprising for three associated reasons:

1. This was a novel tunnelling technique for use in London clay.
2. Success of the method depended upon a systematic relationship between design and construction, the former needing to address all pertinent factors, the latter needing rigorously to conform to the conditions imposed by the former.
3. As explained above, this was a Contract under the terms of the New Engineering Contract allocating powers and duties in an unfamiliar manner between Employer and Contractor, coupled with the responsibility of self-certification by the Contractor as quality control, in the place of independent inspection or assurance of conformity to design.

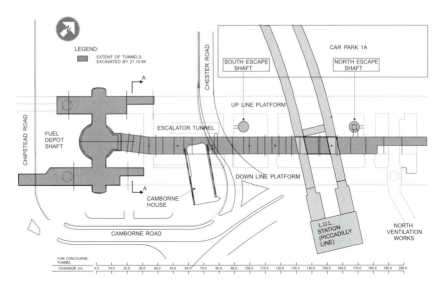

Figure 9.2 Tunnelling at Central Terminal Area (extent of excavation by 20 October 1994).

It is apparent that these issues contributed to the 'climate of risk' (Pugsley 1966) in which the works were being constructed without provision for the counter-measures demanded by such circumstances.

For an unexplained reason, GC had predicted a maximum settlement of only 9 mm as the first tunnel, the Concourse Tunnel, passed beneath Camborne House (Figure 9.2), although comparison with the Trial Tunnel (Atzl and Mayr 1994), would have suggested a figure of around 25 mm – as was in fact predicted by others – in the event, possibly somewhat greater on account of the slower rate of progress of the Concourse Tunnel. Although a shaft was available for use for compensation grouting, mobilisation in advance of tunnelling had not been made for Camborne House. In August 1994, when the Concourse Tunnel was already beyond Camborne House, compensation grouting – or more correctly grout jacking since it was required to lift the ground and the two-storey building – was undertaken in mid-August to reduce settlement, which had already exceeded the figure of 25 mm. This grouting had the effect of depressing the crown of the tunnel by a maximum of 60 mm or more, leading to approximately circumferential cracks around Chainage 60 (measured in metres from axis of the Fuel Depot Shaft). Evidence from optical targets, attached to the periphery of the tunnel (but none was in place between Ch. 54 and 72 – for chainages, see Figure 9.2) for convergence measurement, showed a general depression of the tunnel over a length of at least

72 m. Evidence from the optical targets, as explained in Section 9.3, points to a closure across the tunnel invert. Investigation, which required removal of clay 'running', the clay spoil replaced over the finished invert to a depth of about 1 m to provide a running surface for plant access, indicated that the invert joint had failed, with one side riding over the other. The damaged section was repaired over a length of about 20 m but the repair did not extend to the first constructed length of tunnel from about Ch. 15 to Ch. 54, despite evidence at this time from partial arrays of optical targets at Ch. 30 and 41 of anomalous distortion of the tunnel through this length. In mid-September work started on the lengths of Down Line and Up Line Platform Tunnels, alongside the Concourse Tunnels, spaced for the most part at centres of about 16 m and 21 m from the Concourse Tunnel (Figure 9.3).

Records of movements of the tunnels, obtained from survey by precise theodolite equipped for electronic distance measurement, were usually recorded and plotted daily. Figure 9.4 indicates a vector plot, simplified to show movements of target points (Ch. 41) at monthly intervals. When survey errors are smoothed out, the daily plots at Ch. 30. and Ch. 41 indicate generally continuing distortions at a slightly accelerating rate until the last week of September, when the rate notably increased and the rate of increase continuously accelerated at a period when the Platform Tunnels were approaching Ch. 30. Some concern was expressed by the HEX team but no action was taken. Belatedly, on 18 October,

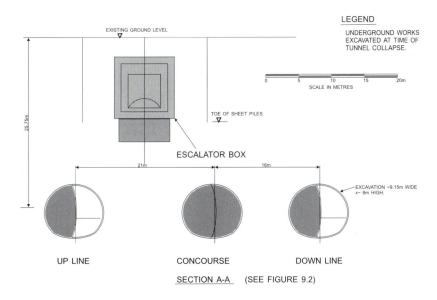

LEGEND

UNDERGROUND WORKS EXCAVATED AT TIME OF TUNNEL COLLAPSE.

EXISTING GROUND LEVEL

0   5   10   15   20m
SCALE IN METRES

25.75m

TOE OF SHEET PILES

ESCALATOR BOX

21m   16m

EXCAVATION ~9.15m WIDE
x~ 8m HIGH.

UP LINE   CONCOURSE   DOWN LINE

SECTION A-A   (SEE FIGURE 9.2)

Figure 9.3 Cross-section of tunnels at Central Terminal Area on 20 October 1994.

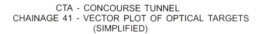

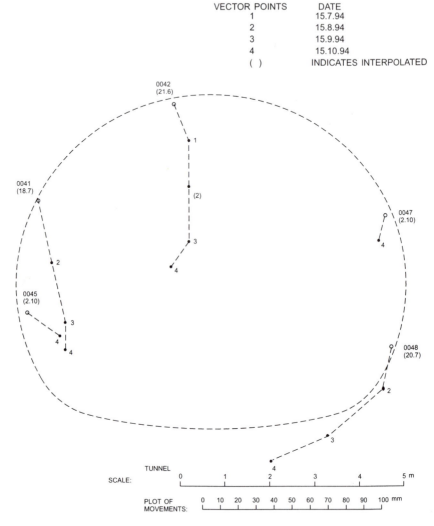

Figure 9.4 Central Terminal Area, Heathrow. Example of partial availability of data for convergence vector plot.

GC admitted to their own concern about the nature of the movements and called for investigation, which started on 19 October. By 20 October, the clay running had been removed from above the central part of the invert over a length of about 20 m, disclosing defects of the joint similar in nature to those already encountered elsewhere in August, described above. Widespread but apparently uncoordinated concrete patching at the invert joint was undertaken. This patching continued to fail until around 02.30 on 20 October 1994, when horizontal cracking of the adjacent side-wall of the Down Line Platform Tunnel, followed by spalling concrete, indicated incipient failure. At around 00.45 on 21 October the tunnels were abandoned and collapse of a length of the Concourse Tunnel from the point of break-out, also the adjacent length of the Down Line Platform Tunnel, occurred around 01.15. During the following two days, further collapses occurred affecting the Up Line Platform Tunnel and an additional length of the Concourse Tunnel in the vicinity of Ch. 54 (affecting part of the length repaired in August 1994). The surface contours post-collapse are affected by surface structures, in particular a slab supported on bored piles over the tunnels extending from the working shaft to approximately Ch. 25. After investigation and grouting of voids in the area affected by the collapses, the project was subsequently completed by sinking a 60 m diameter shaft to envelope much of the collapsed tunnels. The Health and Safety Executive Report on the collapse, 'The Collapse of NATM Tunnels at Heathrow Airport October 1994' is due to be published in 2000.

As a consequence of this chain of events, and the presence of these structures, the pattern of surface subsidence was fairly complicated. There was a general settlement of about 2 metres covering a wide area, with well defined cones of depression, in excess of 3 metres total depth:

(i)   above Ch. 20 of the Down Line Platform Tunnel with a secondary lobe extending to Ch. 30 of the Concourse Tunnel.
(ii) above Ch. 20 of the Up Line Platform Tunnel.
(iii) the deepest and steepest cone at Ch. 60 of the Concourse Tunnel, at the E corner of Camborne House.

The phenomenon of (iii) above was associated with the movements subsequent to the initial collapse. It appears that the crown of the length of tunnel between Ch. 40 and Ch. 54 settled by up to one metre, while beyond Ch. 54 a length of about 10 metres of tunnel collapsed totally. As a consequence, a considerable ingress of clay, gravel and water flowed along the tunnel towards Ch. 80. It seems probable that the lining of the length of tunnel with defective invert folded around itself, causing circumferential cracking at Ch. 54, with the adjacent length of tunnel deformed, but temporarily intact. There would for this reason have been concentrated loading on this length adjacent to Ch. 54 where the discontinuity would provoke rupture, probably initially near the springing of the arch, with progressive failure along this length of tunnel, affected by the previous damage caused by grout-jacking.

9.3 Technical explanation of the collapse

There remains scope for conjecture as to the degree of contribution made by several factors to the collapse. The main purpose of this Chapter is to describe in general terms how and why the collapse occurred and to point to lessons to be learned from this experience. Records of tunnel displacements (e.g. Figure 9.4) provide adequate evidence, notwithstanding sparse data on account of missing targets, of a predominately circumferential component of movement at Ch. 30 and Ch. 41 towards the invert. Mature shotcrete may be assumed to deform linearly elastically, with a modulus of, say, $7–15 \times 10^3$ MPa, depending on quality, up to a working compressive stress of about two-thirds ultimate unconfined compressive strength of around 25 MPa. Thereafter, an increasing degree of creep occurs until, at failure, strain may amount to, say, 0.6–1%. Thus if a lining were subjected to compressive hoop stress approaching failure, strain might amount to 1%, i.e. 40 mm convergence for a radius of 4 m. Other than movement affecting the invert there was no evidence of radial convergence of such a magnitude. The design indicated stress levels of 6–8 MPa to be expected.

At Ch. 30, arrays of stress cells were included, circumferentially between shotcrete lining and ground, also radially within the lining. There are well-known theoretical and practical problems in the use of such cells since their function assumes that they fully represent the stress–strain properties in three dimensions of the concrete they displace. Circumferential cells between ground and structure are difficult to emplace and present particular problems where high shear stress occurs at the interface. Radial cells are buried in shotcrete which itself needs to be carefully applied to avoid 'shadowing' and yet must be representative of the neighbouring shotcrete. There is furthermore the problem of drying shrinkage of the inner surface of the shotcrete which tends to induce bending stress across the thickness of the lining. No results of the circumferential cells have been published. The radial cells show reasonably coherent results with hoop stress usually no more than 2 MPa and with the stress at a position near to target 0035 (i.e. corresponding to the position of 0045 at Ch. 41 on Figure 9.4) falling away to zero from the last week of September 1994. Since the movement of the lining is towards the invert, we might reasonably expect compressive load in the lining to reduce in this direction on account of shear stress between lining and ground.

The amount of foreshortening of the invert could be obtained by summing the circumferential contributions of movements (respectively counter-clockwise and clockwise) of targets such as 0045 and 0048 on Figure 9.4. At Ch. 30 and Ch. 41, on account of late attachment of targets and of failure promptly to replace damaged targets, there are missing data which prevent direct calculation of aggregate movement at any date. Nevertheless, making use of such data which exist, we may establish strain across the invert, the distance between points 0045 and 0048 being about 8 m, from the date of starting readings (which is usually several days after placing the concrete and hence omitting prior movement of young concrete since completion of the ring by 9 July 1994) as Table 9.1, which excludes the accelerating movement during the last few days. For a compressive stress of 2 MPa and assuming a low value of E for the shotcrete of 5×10^3 MPa for such a level of stress, the strain ϵ in the invert should be no more than:

$$\epsilon = 2/(5 \times 10^3) = 0.04\% \tag{9.1}$$

This simple calculation establishes that the invert was incompetent over virtually the full period for which measurements were being taken. If the full arrays had been in position, the defect would have been all the more obvious. While trigger values for convergence may be readily estimated as described below, the cause for *any* persistent indication of circumferential movement of the tunnel shell, once the full ring at any point was completed, should have been investigated.

Although lattice arches had been included around the full periphery of Types (i) and (iii) of the trial tunnel, the redesign for CTA omitted these arches for the invert, the arches for the crown being supported on a bench, the 'elephant's foot'. As a consequence there was no template for the construction of the invert nor any simple positive means for establishing true alignment between the two sides of the invert joint, which was specified to be made while the temporary intermediate wall remained in place. It is certain that errors in construction in the vicinity of this joint represented the main defects in construction of such nature as:

- first (side heading) section of invert cast too flat and too high;
- deficient shotcrete thickness, stated on occasions as less than 50 mm;
- step at junction at foot of side-wall (possibly accentuated by ground movement prior to completing the ring);
- no continuity in reinforcement at junction at foot of temporary side-wall;
- rebound allowed to accumulate, especially near junction.

Table 9.1 Strain across invert of Concourse Tunnel at Ch. 41

Date	Movement (mm)	Strain (%)
15.7.94	from 9.7	?
15.8.94	50	0.6
15.9.94	120	1.5
15.10.94	170	2.1

The failure appears generally to have entailed the first section of invert, that placed as part of the side heading, riding up and forming a step over the section placed to complete the enlargement. A failure of this nature, once started, would tend to be progressive, as developing distortion would lead to transfer of hoop load along the tunnel, and for further stepping and concentration of load across any section of the invert remaining intact.

Simple calculations indicate the order of acceptable convergence of a (near) circular tunnel, without overstress of the shotcrete: (1) *Compressive hoop stress* – a limiting strain of 0.1%, for example, would entail 4 mm radial convergence for a 4 m radius tunnel; (2) *Bending stress* – for a thin elliptical ring of semi-diameters a and b, the radius of curvature at the minor and major axis will be:

$$R_a = b^2/a, \; R_b = a^2/b \tag{9.2}$$

For unchanged length of perimeter, small elliptical distortion of a ring of radius r will give rise to equal and opposite maximum deflections, say $\pm\Delta r$. From eqn (9.2), the radius of curvature at axis will then become respectively:

$$(r - \Delta r)^2/(r + \Delta r) \text{ and } (r + \Delta r)^2/(r - \Delta r), \text{ i.e. } \sim (r \pm 3\Delta r) \tag{9.3}$$

Hence, the change in curvature, $1/\Delta R$, will be given by:

$$1/\Delta R_a \sim [1/(r - 3\Delta r) - 1/r] \text{ and } 1/\Delta R_b \sim [1/(r - 3\Delta r) - 1/r] \tag{9.4}$$

So,

$$1/\Delta R_a = -1/\Delta R_b \sim -3\Delta r/r^2 \tag{9.5}$$

If, for an example, a limiting bending strain, $\epsilon (= \sigma_x/E)$ of 0.1% is assumed, limiting stress σ_x in the concrete at distance from the neutral axis of the ring, z, (say 100 mm) is given by:

$$\sigma_x/z = E/\Delta R_a \text{ or } 1/\Delta R_a = 10^{-5} \text{ mm}^{-1}. \tag{9.6}$$

Thus, from eqns (9.5) and (9.6), for r = 4 m:

$$3\Delta r/(4000)^2 = 1/10^5 \text{ so } \Delta r \sim 50 \text{ mm}. \tag{9.7}$$

Thrusts may then be combined with bending moments, for the known provision of reinforcement, to establish the 'trigger values' for the radial convergence and elliptical distortion to give warning of high stress of the shotcrete, which may be presented for each point as a 'box' (or of one 'box' within another to provide 'amber' and 'red' triggers), in view of the degree of their interdependence.

If, as appears to have been the case for the CTA Concourse Tunnel, the tunnel was wrapping around itself at the invert, it is possible to use an approach similar to that outlined above to determine at what stage cracking and bending failure would be liable to occur. If a reduction ΔC occurs in the circumference, C, with a uniform reduction in radius from r to $(r - \Delta Cr/C)$, i.e. to $(r - \Delta C/2\pi)$, then:

$$1/R = 1/r - 1/(r - \Delta C/2\pi) \sim -\Delta C/2\pi r^2 \qquad (9.8)$$

and, if $1/\Delta R = 10^{-5}$ mm, for $r = 4$ m,

$$\Delta C = 2\pi \times 16 \times 10^6/10^5 \sim 10^3 \text{ mm} \qquad (9.9)$$

indicating a reduction in the tunnel perimeter of about 4%. Such a simple calculation explains why collapse of the Concourse Tunnel occurred only after a massive contraction in the periphery of the tunnel by the failure at the invert. The *mandoria* shape of the heading of the adjacent Down Line Platform Tunnel would be susceptible to failure initially in bending by the release of lateral support combined with increased vertical loading which would have been a result of the reduction in perimeter of the Concourse Tunnel and the associated ground movements downwards and towards this tunnel. The first visible signs of failure of the Platform Tunnel heading would be horizontal cracking, followed by slabbing of the shot-crete – having potential weakness at the interface between layers – and crippling of the steel mesh reinforcement, all features of sketches made by engineers during the incipient period of failure. Collapse of the Down Line Tunnel would then cause failure of the Concourse Tunnel, leading to loss of support of the Up Line Platform Tunnel, which was separated by a greater distance between tunnels (Figure 9.3). Subsequent extension of collapse of the Concourse Tunnel appears to have involved much of the length of tunnel through ground weakened by the degree of shear distortion associated with the compensation grouting of August 1994, and including that section which was repaired in August.

9.4 Failures of management

Having explained, in admittedly over-simplified terms until a comprehensive numerical analysis establishes the timescale for the several contributions to the failure and the phenomena associated with the collapse, it is then necessary, for lessons of value for the future, to understand the circumstances in which these contributory features were permitted to occur.

Firstly, the system of project management was inept for the nature of the work, without strong direction to ensure that the Contractor fully understood the special and unfamiliar responsibilities placed to his charge and assurance that resources and systems were fully adequate in all respects to carry these out. The criteria for success were evidently not understood by BB and the resources committed by GC were quite inadequate to undertake a role of design, monitoring and interpretation of instrumented records of consequences of construction. The duty of ensuring that construction complied with design was a function of GC. Quality control remained with BB without apparent appreciation of the rigorous practices that this should have involved in ensuring compliance with design. There was a procedure of 'Corrective Action Requests' which needed to be 'signed off' by GC, but subject to a process whereby such notices would not be seen until an elapse of days, by which time any defect in the invert for example would be covered up and any remedial work would have caused major disruption to progress. (After the events of August 1994, the practices were improved which may account for the failure stopping short of the traverse beneath the Piccadilly Line, Figure 9.2.)

GC was represented at site by a single engineer experienced in NATM, who had also to cover similar work at Terminal 4 and, for a period, junction lengths at Skepiston Lane, Sipsons Farm and Custom House shafts (Figure 9.1). BB employed three, later four, so-called NATM engineers whose principal function was the installation and reading of instruments (but not of the vital optical targets on the lining for measuring 'convergence' which was undertaken by BB surveyors). They had no specific responsibility for the quality of construction work although they commented upon it from time to time. The arrays of optical targets were not adequately maintained but there is no evidence of any insistence upon ensuring that all targets were in place or replaced following damage, although the widely circulated records must have indicated the extent of deficiencies. All those familiar with this type of work emphasise that these indicators of 'convergence' provide the most important and reliable source of monitoring data.

The general picture is that of failure of BB to recognise their vital role in a design-led system of construction. GC should have committed resources adequate for their responsibilities; they should also have concentrated upon the most vital priority, that of monitoring tunnel movements against established 'trigger values' which required explanation or investigation. The specific feature of excessive circumferential movements of tunnel optical targets should have given rise to a requirement for immediate investigation which would have resulted in identification of the faulty workmanship. It is particularly remarkable that the tunnel invert for the initial length of the Concourse Tunnel up to about Ch. 54 was not investigated following the failure associated with grout-jacking in

August 1994 described earlier since this was the length of tunnel at the beginning of the learning curve.

The question needs to be asked as to why, if the evidence for incipient collapse was so obvious, none of the engineers on the project, to whom the records were circulated at intervals, asked searching questions requiring explanations of the anomalous behaviour of the tunnel (the tentative questions that were asked were given bland reassuring replies). This feature is indeed difficult to explain but at least a partial explanation is found in the air of esoteric mystery that pervades NATM, encouraged by many of its promoters. Those who assume that there indeed exists an occult art, beyond the rules for good engineering, may put unjustifiable faith in the 'experts' with the undesirable effect that scepticism is set aside and even the most obvious questions not asked. A more normal practice among engineers, of explaining what they are doing and why, would have avoided development of such a degree of unwarranted confidence. It must now be apparent that all that was required from GC in monitoring construction was to exercise straight-forward application of sound engineering principles.

9.5 Summary of factors contributing to failure

Summarised below are a number of factors seen as contributing to the collapse of the Heathrow tunnels. These features appear sufficient to explain the nature and timing of the collapse without the need to conjure up a *deus ex machina*. What is remarkable is that if any one of the issues set out below had been addressed competently, in all probability the collapse would not have occurred. Although no lives were lost, this was an expensive accident. The cost of the collapse is stated (*New Civil Engineer* 18 February 1999 p. 4) as £35M to BB, £35M to BAA, £100M to BAA and BB insurers, £50M for airport disruption, £200M for the associated stoppage of working on NATM construction for the Jubilee Line Extension, apart from court fines of £1.2M to BB and £0.5M to GC plus costs (appeal by GC was subsequently dismissed). The original tender value was around £60M. Other consequential costs are excluded from these figures.

9.5.1 The project management

The underlying cause for the incident is undoubtedly to be found in the adoption of an unfamiliar system of project management based on the New Engineering Contract (NEC) (Institution of Civil Engineers 1993), without thought for the special measures necessary to ensure that the responsibilities placed on the Contractor were fully specified, recognised and implemented. The New Engineering Contract is essentially a framework document, with different options for the contractual base, which requires to be accompanied by a procurement strategy taking account of the special features of the project.

9.5.2 Relationship between design and construction

On the one hand, adoption of the method of construction was predicated on special knowledge on design and monitoring to be brought by GC; on the other hand, pressures towards least cost confined GC to resources inadequate to provide even minimal control of the work, and lacking the powers of control appropriate for a design-led system of construction. It should have been evident from the outset that GC were unable, on account of limits on their powers and resources, to undertake the role on which success depended. With the function of self-certification imposed on BB, it was essential, at the time of tender, to ensure a fully adequate system of design control of this work since additional requirements imposed subsequently would lead to increased costs to the Employer.

The NEC contains no advice on the exceptions to the statement of a break in the design process at the time of inviting tenders (Section 7.7). Where it is expedient to procure a project through design-and-build, there is bound to be a hiatus at this stage. Where, as for tunnelling, it is essential to overcome this problem, it is the job of the drafter of the Contract terms to make provision, in one of several possible means, to overcome this problem, possibly by a form of 'partnering', prior to negotiation on terms, and in any circumstances ensuring that the design concepts and assumptions are fully understood and taken into account by the Contractor. Quality assurance for many forms of tunnelling will need to address the specific needs for a tunnel to be 'right first time' and to preclude vital defects from being overlooked. Here once again (and see Chapter 7) it is necessary to ensure that QA procedures are not confined to a narrow definition of permanent works, in recognition that phases of construction procedures may be vital to the temporary – or even longer term – security of the works. Broader criteria for good practice are described elsewhere in this book.

9.5.3 Acceptance standards of construction

The contract placed on BB the responsibility that construction complied with design. Only after completion of a considerable length of tunnelling was there counter-action following appreciation that the system of Corrective Action Requests (CARs) and System Defect Notices (SDNs) was not ensuring full compliance with the requirements of design. This did not however lead to investigation of the extent of hidden defects of the section already built which would have allowed a planned correction programme in a timely manner.

9.5.4 Compensation grouting

As described in Section 9.3, compensation grouting under Camborne House was in reality grout-jacking, requiring pressures in excess of pre-existing vertical ground stress. The ratio of horizontal to vertical stress (K_0) in the London clay at Heathrow at the depth of the tunnels is probably not greatly in excess of unity (assumed by different authors for purposes of analysis as between 1.15 and 1.5). The prior construction of the tunnel would have been associated with ground movements towards the tunnel which would have further reduced local horizontal stress, probably also affected by the failure of the tunnel invert prior to this time. In consequence, the excess pressure at the grouting point could well have exceeded the *in situ* horizontal ground stress, which would cause grout to follow joints or paths of weakness at inclined or even sub-vertical angles. This could then have added to the loading on the crippled tunnel, contributing to the already noted tunnel distortion. Evidence for such inclined grout paths was disclosed by the excavation subsequent to the collapse.

9.5.5 Monitoring

There is no evidence that there was any reaction by GC to the consistent records of circumferential movement of the tunnel lining coupled with depression of the crown. There was no acceptable explanation for this phenomenon other than a weakness of the invert. The fact that such weakness had been exposed over the length of invert investigated in August 1994 makes this lack of reaction all the more extraordinary. Since GC did not indicate to others any hint of the magnitudes of acceptable deflections or movement of the lining, there was no quantitative 'trigger' to spur calls for action from others. The poor maintenance of the arrays of optical targets would have made it appear that there was no serious cause for such monitoring to be undertaken beyond the provision of data, which data were provided in figures and diagrams devoid of comment. Nevertheless, the data caused BAA to question the integrity of the tunnel, an apprehension dismissed summarily by BB. Here, again, the over-dependence on an engineer with claims for exclusive skill, is exposed as a hazard. The acceptance of the evidence of incipient failure did not require egregious skill.

It has been suggested that acceleration of recorded closure of the tunnel invert during the last fortnight prior to the collapse might have been fully attributable to the advance and temporary stoppage of the adjacent Platform Tunnels. This suggestion does not bear examination since Ch. 41 is well beyond any affective zone of influence of the partially excavated length of the Platform Tunnels, so long as they remained intact, prior to the collapse.

9.5.6 *Failure to investigate*

In many forms of contract, there is a requirement on the Contractor to investigate any suspected defect, for which the Contractor is reimbursed if no such defect is found. There was no similar imposition on BB so that, in the event of a rebuttal by BB of an expression of concern by HEX, a major 'compensation event' (in the terms of the NEC) would have arisen. The operation of audit by HEX was inadequate to establish the extent to which the process of 'self-certification' was failing.

9.6 Events post collapse

After the causes for collapse had been investigated, work was resumed with major changes in procedure (Powell *et al.* 1997). MM responsibilities for design were extended to include Informal Support work, the basis of the Contract with BB became a Target Contract, and steps were taken towards 'partnering' in the resolution of problems – all generally features of a late blossoming of enlightenment.

References

Allport, R.J. and Von Einsiedel, N. (1986) An innovative approach to metropolitan management in the Philippines. *Pub. Admin. Dev.*, 6(1), 23–48.

Altounyan, P.F.R and Farmer, I.W. (1981) Tunnel lining pressures during groundwater freezing and thawing, in *Proc. 5th Rapid Exc. Tunnelling Conf.*, San Francisco Society of Mining Engineers of AIME, San Francisco, pp. 784–800.

Anagnostou, G. and Kovári, K. (1996) Face stability in slurry and EPB shield tunnelling, in Mair and Taylor (eds.) 1996, pp. 453–8.

Andrews, K.E., McIntyre, B.E. and Mattner, R.H. (1964) Some aspects of high speed rock tunnelling in the Snowy Mountains. *Civ. Eng. Trans. Inst. Engrs, Australia*, September, paper no. 1781, CE6, 51–70.

Anon. (1974) Kariba machine hall, *New Civil Engr.*, 18 April, 6–13.

Anon. (1977) *Guidelines for Professional Performance Audit* (unpublished), Sir William Halcrow and Partners, London.

Anon. (1998) Solving problems on the Hollandsäs Project, *Tunnels Tunnelling*, May, 22–4.

ASCE (1997) *Geotechnical Baseline Report*. The Technical Committee on Geotechnical Reports of the Underground Technology Research Council (ed. R. Essex), ASCE, London.

Attewell, P.B., Yeates, J. and Selby, A.R. (1986) *Soil Movements Induced by Tunnelling and Their Effects on Pipelines and Structures*, Blackie, London.

Atzl, G.V. and Mayr, J.K. (1994) FEM-analysis of Heathrow NATM trial tunnel, in *Numerical Methods in Geotechnical Engineering*, Balkema, Rotterdam, pp. 195–201.

Barthes, H. *et al.* (1994) The Channel Tunnel, *Proc. Inst. Civ. Engrs Channel Tunnel Supp. to Civ. Engng*, 102, Special Issue 1, Part 3, 2–88.

Bartlett, J.V., Biggart, A.R. and Triggs, R.L. (1973) The bentonite tunnelling machine with thixotropic slurry support of the face, *Proc. Inst. Civ. Engrs*, 54, 605–24.

Bazalgette, J.W. (1865) On the main drainage of London, *Min. Proc. Inst. Civ. Engrs*, 24, 280–358.

Bebi, P.C. and Mettier, K.R. (1979) Ground freezing for the construction of the three-lane Milchbuck road tunnel, Zurich, in *Tunnelling '79*. Institute of Mining and Metallurgy, London.

Berry, F.G. (1979) Late Quaternary scour-hollows and related features in Central London, *Q. J. Engng Geol.*, 12, 9–29.

Biggart, A.R. and Sternath, R. (1996) Storebaelt eastern railway tunnel: construction, *Proc. Inst. Civ. Engrs. Civ. Engng Supp.*, **114**(1), 20–39.

Blockley, D.I. (ed.) (1992) *Engineering Safety*, McGraw-Hill, New York.

Boscardin, M.D. and Cording, E.J. (1989) Building response to excavation induced settlement, *J. Geotech. Engng ASCE*, **115**, 1–21.

Bougard, J.F. (1988) The mechanical precutting method. *Tunnelling Undergr. Space Technol.*, **3**(2), 163–7.

Bowers, K.H., Hiller, D.M. and New, B.M. (1996) Ground movement over three years at the Heathrow Express Trial Tunnel, in Mair and Taylor (eds.) 1996, pp. 647–52.

Bracegirdle, A., Mair, R.J., Nyren, R.J. and Taylor, R.N. (1996) A methodology for evaluating potential damage to cast iron pipes induced by tunnelling, in Mair and Taylor (eds.) 1996, pp. 659–64.

British Standards Institution (1995) *Eurocode 7: Geotechnical Design. Part 1: General Rules. Development for Draft DD ENV 1997–1:1995.*

Broch, E. and Nilsen, B. (1990) Hard-rock tunnelling, in Edwards (ed.) 1990, pp. 387–413.

Broms, B.B. and Bennermark, H. (1967) Stability of clay at vertical openings, *J. Soil Mech. Found. Engng ASCE*, **93** SM1, 71–4.

Brown, E.T. (1981) Putting the NATM into perspective, *Tunnels and Tunnelling*, November, 13–17.

Bruckshaw, J.M., Goguel, J., Harding, H.J.B. and Malcor, R. (1961) The work of the Channel Tunnel Study Group 1958–60. *Proc. Inst. Civ. Engrs*, **18**, 149–78.

Bruland, A. (1998) *Prediction Model for Performance and Cost of Norwegian TBM Tunnelling.* Norwegian Soil and Rock Engineering Association Publication II, 29–34.

Brunel, M.I. (1818) *Specification for Patent Application No. 4204 Forming Tunnels or Drifts Underground*, Great Seal Patent Office, London.

Brux, G. (1997) Channel Tunnel fire, *Tunnel (Köln)*, **6/97**, 31–42.

BS 5930 (1981) *British Standard Code of Practice for Site Investigation*, British Standards Institution, London.

BS 4778 (1991) *Quality Vocabulary*, British Standards Institution, London.

Bulson, P.S. (1985) *Buried Structures*, Chapman & Hall, London.

Burland, J.B. and Wroth, C.P. (1974) Settlement of buildings and associated damage, *Proc. Conf. on Settlement of Structures*, Cambridge, Pentech Press, London, pp. 611–54.

Cambefort, H. (1977) The principles and applications of grouting, *Q. J. Engng. Geol.*, **10**, 57–95.

Campion, F.E. (1967) Part reconstruction of Bo-Peep tunnel at St Leonards-on-Sea, *J. Inst. Civ. Engrs*, **5**, 52–75.

Cartwright, D.E. and Crease. J. (1963) A comparison of the geodetic reference levels of England and France by means of sea surface, *Proc. Roy. Soc. (A)*, **273**, 558–80.

Chaplin, R. (1989) *Creativity in Engineering Design*, Fellowship of Engineering (later Royal Academy of Engineering), London.

Chapeau, C. and Dupuy, D. (1976) Application du procédé de congélation à l'azote liquide, *Tunnels et Ouvrages Souterrains*, **16**, 143–51.

CIRIA (1978) *Tunnelling – Improved Contract Practices*, Construction Industry Research and Information Association Report 79, CIRIA, London.

CIRIA (1979) *Tunnel Waterproofing*, Construction Industry Research and Information Association Report 81, CIRIA, London.

CIRIA (1997) *The Observational Method in Ground Engineering: Principles and Applications*, Founders Report/CP/49, CIRIA, London.

Clark, J.A.M., Hook, G.S., Lee, J.J. Mason, P.L. and Thomas, D.G. (1969) Victoria Line: some modern developments in tunnelling construction, *Proc. Inst. Civ. Engrs. (Supp.)*, 397–451.

Coats, D.J., Berry, N.S.M. and Banks, D.J. (1982) The Keilder Transfer Works, *Proc. Inst. Civ. Engrs*, **72**(1), 177–208.

CONIAC (1994) *Designing for Health and Safety*, Health and Safety Commission, Construction Industry Advisory Committee, London.

Copperthwaite, W.C. (1906) *Tunnelling Shields and the Use of Compressed Air in Subaqueous Works*. Constable, London.

Craig, R.N. (1979) The Lewes Tunnel, Sussex, England, *Tunnelling '79*, 153–7.

Craig, R.N. and Muir Wood, A.M. (1978) *A Review of Tunnel Lining Practice in the United Kingdom*, Transport and Road Research Laboratory Supplementary Report 335, TRRL, Crowthorne, UK.

Curtis, D.J. (1976) Correspondence on Muir Wood (1975a), *Géotechnique*, **26**, 231–7.

Cuthbert, E.W. and Wood, F. (1962) The Thames to Lee tunnel water main, *Proc. Inst. Civ. Engrs*, **21**, 257–76.

Dalrymple-Hay, H.H. (1899) The Waterloo and City Railway, *Min. Proc. Inst. Civ. Engrs*, **139**, 25–55.

Dann, H.E., Harting, W.P. and Hunter, J.R. (1964) Unlined tunnels of the Snowy Mountains Hydro-Electric Authority, *ASCE J. Power Div.*, PO3 **90**, 47–79.

DAUB (1997) (Deutscher Ausschuss für unterirdisches Bauen) Recommendataions for selecting and evaluating Tunnel Boring Machines, *Tunnel, (Köln)*, **5/97**, 20–35.

Davis, E.H., Gunn, M.J., Mair, R.J. and Seneviratne, H.N. (1980). Stability of shallow tunnels and underground openings in cohesive material. *Géotechnique*, **30**(4), 397–416.

Deere, D. *et al.* (1967) Design of surface and near surface construction in rock, *Proc. 8th Symp. Rock Mechanics*, Minnesota, American Institute of Mining Engineering, Minnesota.

Donovan, H.J. (1969) Modern tunnelling methods, *J. Inst. Public Health Engrs*, **68**(2), 103–39.

Duddeck, H. (1981) Safety at the design stage of a tunnel, *Adv. Tunnel Technol. Subsurface Use*, **1**(2), 87–91.

Duddeck, H. and Erdmann, J. (1982) Structural design models for tunnels, *Tunnelling '82*, Institute of Mining and Metallurgy London, pp. 83–92.

Dunton, C.E., Kell, J. and Morgan, H.D. (1965) Victoria Line: experimentation, design, programme and early progress, *Proc. Inst. Civ. Engrs*, **31**, 1–24.

Edwards, J.T. (ed) (1990) *Civil Engineering for Underground Rail Transport*, Butterworth, London.

Egger, P. (1996) Tunnel construction in Stuttgart: problems of settlement and swelling rock, in Mair and Taylor (eds) 1996, pp. 263–8.

Elliott, I.H., Odgård, A.S. and Curtis, D.J. (1996) Storebaelt eastern railway tunnel: design, *Proc. Inst. Civ. Engrs Civ Engng Supp.*, **114**(1), 9–19.

Eves, R.C.W. and Curtis, D.J. (1992) Tunnel lining design and procurement, *Proc. Inst. Civ. Engrs Civ. Engng Channel Tunnel*, **1**, 127–43.

Fechtig, R. and Kovári, K. (1996) *Historische Alpendurchstiche in der Schweiz*, ETH, Zurich.

FIDIC (1987) *Conditions of Contract for Works of Civil Engineering Construction*, 4th edn, Fédération Internationale des Ingénieurs-Conseils, Laussane.

Flury, S. and Rehbock-Sandes, M. (1998) The Gotthard Base Tunnel. Stage reached in planning and construction, *Tunnel (Köln)*, **4/98**, 26–31.

Follenfant, H.G. *et al.* (1969) The Victoria Line, *Proc. Inst. Civ. Engrs (Supp.)*, Supplementary volume, 337–477.

Fugeman, I.C.D., Hawley, J. and Myers, A.G. (1992) Major underground structures, *Proc. Inst. Civ. Engrs Civ. Engng*, Channel Tunnel Part 1: Tunnels, 87–102.

Geological Society of London (1970) The logging of rock cores for engineering purposes, *Q. J. Engng Geol.*, **3**, 1–25.

Greater London Council (1973) *Roads in Tunnels*, Greater London Council, London.

Glossop, R. (1968) The rise of geotechnology and its influence on engineering practice, *Géotechnique*, **18**(2), 107–50.

Godfrey, P.S. (1996) *Control of Risk*, CIRIA Special Pub. 125, CIRA, London.

Greathead, J.H. (1896) The City and South London Railway, *Proc. Inst. Civ. Engrs*, **123**, 39–73.

Groves, G.L. (1943) Tunnel linings, with special reference to a new form of reinforced concrete lining, *J. Inst. Civ. Engrs*, **20**, 29–41.

Gysel, M. (1977) A contribution to the design of a tunnel lining in swelling rock, *Rock Mech.*, **10**, 55–71.

Hackel, A. (1997) Facing the Piora trough, *World Tunnelling*, November, 408–10.

Harding, H.J.B. (1946) The choice of expedients in civil engineering construction, *Proc. Inst. Civ. Engrs D(Wks)* **6**, 3–30.

Harding, H.J.B. (1981) *Tunnelling History and My Own Involvement*. Golder Assoc., Toronto.

Harris, C.S. *et al.* (eds.) (1996) *Engineering Geology of the Channel Tunnel*, Thomas Telford, London.

Harris, C.S., Hart, M.B. and Wood, C.J. (1996) A revised stratigraphy, in Harris *et al.* (eds) 1996, pp. 398–420.

Harvey, W.J. (1988) Application of the mechanism analysis to masonry arches, *Struc. Engr.*, **66**(5), 77–84.

Haswell, C.K. (1973) Tunnel under the Severn and Wye estuaries, *Proc. Inst. Civ. Engrs*, **52**, August, 451–86.

Hay, D. and Fitzmaurice, M. (1897) The Blackwell Tunnel, *Min. Proc. Inst. Civ. Engrs*, **130**, 50–79.

Health and Safety Commission (1991) *Major Hazard Aspects of Transport of Dangerous Substances*, HMSO, London.

Health and Safety Executive (1985) *The Abbeystead Explosion – A Report of the Investigation by the Health and Safety Executive into the Explosion on 23 May 1984 at the Valve House of the Lune/Wyre water Transfer Scheme at Abbeystead*, HMSO, London.

Health and Safety Executive (1996) *Safety of New Austrian Tunnelling Method (NATM) Tunnels*, HMSO, London.

Hentschel, H. (1997) Constructing the immersed tunnel beneath the Øresund, *Tunnel (Köln)*, **7/97**, 32–42.

Heyman, J. (1982) *The Masonry Arch*, Ellis Horwood, Chichester.

Higgins, K.G., Mair, R.J. and Potts, D.M. (1996) Numerical modelling of the influence of Westminster Station excavation and tunnelling on the Big Ben Tower, in Mair and Taylor (eds) 1996.

Hillier, D.M. and Bowers, K.H. (1997) Ground-borne vibration from mechanical tunnelling works, *Tunnelling '97*, 721–35.

Hobbs, N.B. (1974) Factors affecting the prediction of settlement of structures on rock: with particular reference to Chalk and Trias, in *Proc. Conf. Settlement of Structures*, British Geotechnical Society, Cambridge, pp. 579–610.

Hoek, E. (1983) Strength of jointed rock masses, *Géotechnique*, 33(3), 187–223.

Hoek, E. and Brown, E.T. (1980) *Underground Excavation in Rock*, Institute of Mining and Metallurgy, London.

Hudson, J.A. (1993) The construction process, in *Comprehensive Rock Engineering*, vol. 4, (ed. J.A.Hudson), Pergamon, Oxford, pp. 1–37.

Humphrey, J.W., Oleson, J.P. and Sherwood, A.N. (1998) *Greek and Roman Technology*, Routledge London.

Hutchinson, J.N. (1991) Theme Lecture – Periglacial and slope processes, *Geol. Soc. Lond. Engng Geol. Spec. Pub.*, 7, 283–331.

Institution of Chemical Engineers (1992) *IChemE Model Form of Conditions of Contract for Process Plant, Reimbursable Contract (2nd edn)*, Institution of Chemical Engineers, Rugby.

Institution of Civil Engineers (1969) *Engineering Economics*, Thomas Telford, London.

Institution of Civil Engineers (1991) *ICE Conditions of Contract* (6th Ed), Thomas Telford, London.

Institution of Civil Engineers (1993) *The New Engineering Contract*, Thomas Telford, London.

Institution of Civil Engineers (1995) *The NEC Engineering and Construction Contract*, Thomas Telford, London.

Institution of Civil Engineers (1996) *Sustainability and Acceptability in Infrastructure Development*, Thomas Telford, London.

Institution of Civil Engineers (1997) *Sprayed Concrete Linings (NATM) for Tunnels in Soft Ground*, Thomas Telford, London.

Institution of Civil Engineers, Faculty and Institute of Actuaries (1998) *Risk Analysis and Management of Projects*, Thomas Telford, London.

International Tunnelling Association (1991) *Shotcrete in Tunnelling*, Swedish Rock Engineering Research Foundation, Stockholm.

Jacobs, C.M. (1910) The Hudson River tunnels of the Hudson and Manhattan Railroad Company, *Min Proc. Inst. Civ. Engrs*, 181, 169–221.

Jaeger, C. (1955) Present trends in the design of large pressure tunnels and shafts for underground hydro-electric power stations, 4(1), 116–200.

Jaeger, J.C. and Cook, N.G.W. (1979) *Fundamentals of Rock Mechanics*, Proc. Inst. Civ. Engrs, 3rd edn, Chapman & Hall, London.

Jobling, D.G. and Lyons, A.C. (1976) Extension of the Piccadilly Line from Hounslow to Heathrow Central, *Proc. Inst. Civ.Engrs*, **60**, 191–218.

John, M. (1980) Construction of the Arlberg expressway tunnel tube, *Tunnels and Tunnelling*, **12**(5), 45–50 and (6), 66–8.

John, M. and Allen, R. (1996) NATM on the Channel Tunnel, in Harris *et al.* (eds) 1996, pp. 310–36.

Jones, M. (1998) Docklands Light Railway beneath Thames from Island Gardens to Cutty Sark Station, *New Civ. Engr*, 26 February, 3–5.

Jones, R.H. (1996) Observation and control of movement in works constructed by ground freezing. *Geotechnical Applications of Underground Construction in Soft Ground*, Balkema, Rotterdam, pp. 379–84.

Jøsang, T. (1980) Ground freezing techniques used for tunnelling in Oslo city centre, in *Proc. 2nd Int. Symp. on Ground Freezing*, Tapir, Trondheim.

Kasten, H.A.D. (1997) Fibre reinforced shotcrete, *World Tunnelling*, **10**(9), 411–15.

Kastner, H. (1971) *Statik des Tunnels und Stollenbaues*, Springer-Verlag, Berlin.

Kell, J. (1963) The Dartford Tunnel, *Proc. Inst. Civ. Engrs*, **24**, 359–72.

Kell, J. and Ridley, G. (1966) Blackwall Tunnel Duplication, *Proc. Inst. Civ. Engrs*, **35**, 253–74.

Kerisel, J. (1969) Underground town planning – informal discussion, *Proc. Inst. Civ. Engrs*, **43**, 109–111.

Kidd, B.C. (1976). Instrumentation of underground civil structures, *Proc. Symp. on Exploration for Rock Engineering*, Johannesburg, pp. 210–31.

Kolymbas, D. (1998) *Geotechnik – Tunnelbau und Tunnelmechanik*, Springer-Verlag, Berlin.

Kommerell, O. (1940) *Statische Berechnung von Tunnel mauwerk*, Ernst & Sohn, Berlin.

Kovári, K. (1993) *Erroneous Concepts Behind NATM*, ETH, Zurich.

Kovári, K. (1998) Tunnelling in squeezing rock, *Tunnel (Köln)*, **5/98**, 12–31.

Kovári, K. (1999) Tunnelling in squeezing rock: a challenge for the New Alpine Traversals. *Proc. Conf. Int. Soc. Rock. Mech.* (Paris), in press.

Ladanyi, B. (1974) Use of the long-term strength concept in the determination of ground pressure on tunnel linings, in *Adv. Rock Mechs, Proc. 3rd Int. Conf. Int. Soc. Rock Mech.*, **II**(B), pp. 1150–6.

Lane, K.S. (1975) Role of test sections for reducing costs in tunnel support, *Project 211.03*, ASCE.

Lang, T.A. (1961) Theory and practice of rock-bolting, *Trans. Am. Inst. Min. Metall. Pet. Engng*, **220**, 333–48.

Latham, M. (1994) *Constructing the Team: Final Report of the Government/ Industry Review of Procurement and Contractual Arrangements in the UK Construction Industry*, HMSO, London.

Law, H. (1845) A memoir of the Thames tunnel, reprinted from *Quarterly Papers in Engineering*, volume 3.

Leblais, Y. and Leblond, L. (1996) French undersea crossover; design and construction, in Harris *et al.* (eds) 1996, pp. 349–56.

Legget, R.F. (1979) Geology and geotechnical engineering, *J. Geotech. Engng. Div. Proc. ASCE*, **105**, GT3, 342–91.

Legget, R.F. and Hatheway, A.W. (1988) *Geology and Engineering*, McGraw-Hill, New York.

Lemoine, B. (1991) *Le Tunnel sous la Manche*, Editions du Moniteur, Paris.

Lichtsteiner, F. (1997) Grouting measures in practice; a survey of injection systems, *Tunnel (Köln)*, **7/97**, 43–50.

Lu, Z.K. and Wrobel, L.C. (1997) Analysis of the high pressure tunnel of the Guangzhou pumped storage power tunnel, *Proc. Inst. Civ. Engrs. Marit. Energy*, **124**(1), 25–31.

Lunardi, P. (1997) The influence of the rigidity of the advance core on the safety of the tunnel excavation, *Gallerie*, **52**, 16–37 and *Tunnel (Köln)*, **8/98**, 32–44.

Lunardi, P. (1998) The Bologna to Florence high speed rail connection. Design and construction aspects of the underground works, *Gallerie*, **54**, 20–31.

Lunardi, P., Focaracci, A. and Merlo, S. (1997) Mechanical precutting for the construction of the 21.5 m span arch of the 'Baldo degli Ubaldi' station, *Gallerie*, **53**, 54–69.

Lyra, F.H. and MacGregor, W. (1967) Furnas hydro-electric scheme, Brazil; closure of diversion tunnels, *Proc. Inst. Civ. Engrs*, **36**, 20–46.

Mair, R.J. (1981). The prediction and safety of tunnels under construction in soft ground by centrifuge model testing, *Adv. Tunnel Technol. Subsurface Use*, **1**(2), 113–18.

Mair, R.J. and Taylor, R.N. (eds) (1996) *Proc. Int. Symp.on Geotechnical Aspects of Underground Construction in Soft Ground*, Balkema, Rotterdam.

Mair, R.J., Taylor, R.N. and Bracegirdle, A. (1993) Subsurface settlement profile above tunnels in clays, *Géotechnique*, **43**(2), 315–20.

Mair, R.J., Taylor, R.N. and Burland, J.B. (1996) Prediction of ground movements and assessment of risk of building damage due to bored tunnelling, in Mair and Taylor 1996, pp. 713–18.

Marchini, H. (1990) Ground treatment, in *Civil Engineering for Underground Rail Transport*, J.T. Edwards (ed.), Butterworths, London, pp. 57–86.

Morgan, H.D., Haswell, C.K. and Pirie, E.S. (1965) The Clyde Tunnel: Design, construction and tunnel services, *Proc. Inst Civ. Engrs*, **30**, 291–322.

Mortimore, R.N. and Pomerol, B. (1996) Chalk Marl: geoframeworks and engineering, in Harris *et al.* (eds) 1996, pp. 455–66.

Moy, D. (1995) A review of the major rock classification schemes applicable to tunnel and cavern support design, in *Conf. on Design and Construction of Underground Structures*, Assoc. Consultg. Civil Engrs and Indian Soc. For Roch Mech. And Tunnelling Technology (New Delhi), 21–41.

Moye, D.G. (1965) Unstable rock and its treatment in underground works in the Snowy Mountains scheme, in *Proc. 8th Commonwealth Mining and Metallurgy Cong.*, **6**, 429–44.

Muir Wood, A.M. (1970) Soft ground tunnelling, *Technol Pot. Tunnelling* (Johannesburg), **1**, 167–74; **2**, 72–5.

Muir Wood, A.M. (1972) Tunnels for roads and motorways, *Q. J. Engng. Geol.*, **5**(1, 2), 111–26.

Muir Wood, A.M. (1975a) The circular tunnel in elastic ground, *Géotechnique*, **25**(1), 115–27.

Muir Wood, A.M. (1975b) Tunnel hazards: UK experience, *Hazards in Tunnelling and on Falsework*, Institution of Civil Engineers, London, pp. 149–64.

Muir Wood, A.M. (1976) A century of British tunnelling, *Tunnels Tunnelling*, November, 53–8.

Muir Wood, A.M. (1978) Presidential Address to Institution of Civil Engineers, *Proc. Inst. Civ. Engrs*, **64**(1), 1–23.

Muir Wood, A.M. (1987) To NATM or not to NATM, *Felsbau*, **5**(1), 26–30.

Muir Wood, A.M. (1993) Development of tunnel support philosophy, *Comprehensive Rock Engineering*, Vol. 4, (ed. J.A.Hudson), Pergamon, Oxford, pp. 349–68.

Muir Wood, A.M. (1994a) The Thames Tunnel 1825–43: where shield tunnelling began, *Proc. Inst. Civ. Engrs Civ. Engng*, **102**, 130–9.

Muir Wood, A.M. (1994b) Will the newcomer stand up? *Tunnels Tunnelling*, September, 30–2.

Muir Wood, A.M. (1995). *The First Road Tunnel*, Permanent International Association of Road Congresses (PIARC), Paris.

Muir Wood, A.M. (1996) Systems engineering and tunnelling. A new look, *Gallerie*, **50**, 16–24.

Muir Wood, A.M. and Casté, G. (1970) Insitu testing for the Channel Tunnel, *Proc. Conf. Insitu Investigations in Soils and Rocks*, British Geotech. Soc. Inst. Civ. Engrs, London, pp. 79–85.

Muir Wood, A.M. and Duffy, F. (1996) Realising our potential – in the built environment. *Education for the Built Environment*, Ove Arup Foundation, London.

Muir Wood, A.M. and Gibb, F.R. (1971) Design and construction of the Cargo Tunnel at Heathrow Airport, London, *Proc. Inst. Civ. Engrs*, **48**(2), 11–34.

Müller, L. (1978) *Der Felsbau*, Enke, Salzburg.

Müller, L. and Fecker, E. (1978) Grundgedanken und Grundsätze der 'Neuen Österreichischen Tunnelbauweise', *Grundlagen und Anwendung der Felsmechanik*, Karlsruhe, 247–62.

Munro, M. (1997) Hong Kong setback, *World Tunnelling*, **10**(6), 250–4.

Needham, J. (1971) *Scie. Civil. China*, 4(3), section 28, Cambridge University Press, Cambridge.

Neumann, J. von and Morgenstern, G. (1944). *The Theory of Games and Economic Behavior*, Princeton University Press, Princeton.

O'Reilly, M.P. and New, B.M. (1982) Settlement above tunnels in the United Kingdom – their magnitude and prediction, *Tunnelling '82*, Institute of Mining and Metallurgy, London, pp. 137–81.

Orr, W.E., Muir Wood, A.M., Beaver, J.L., Ireland, R.J. and Beagley, D.P. (1991) Abbeystead Outfall Works: background to repairs and modifications – and lessons learned, *J. Inst. Wat. Envir. Man.*, **5**(1), 7–20.

Panet, M. and Guenot, A. (1982) Analysis of convergence behind the face of a tunnel, *Tunnelling '82*, Institute of Mining and Metallurgy, London, pp. 197–204.

Patrucco, M. (1997) Mechanized tunnel driving machines: the significance and evolution of European Community technical standards, *Gallerie*, **52**, 52–60.

Peck, R.B. (1969a) Deep excavation and tunneling in soft ground – State of the Art Volume, *Proc. 7th Int. Conf. Soil Mechanics and Foundation Engineering*, Mexico City, Inst. Soc. Soil Mechanics and Foundation Engineers, Mexico City, pp. 225–90.

Peck, R.B. (1969b) Advantages and limitations of the observational method in applied soil mechanics, *Géotechnique*, **19**(2), 171–87.

Pelizza, S. and Grasso, P. (1998) Tunnel collapses, *World Tunnelling*, **11**(2), 71–6.

Plichon, J.N. (1974) Le tunnel d'Eupalinos, *Tunnels et Ouvrages Souterrains*, 5, 205–9.

Pollard, C., Green, T.J. and Conway, R.G. (1992). Construction planning and logistics, *Proc. Inst. Civ. Engrs Civ. Engng Channel Tunnel*, 1, 103–26.

Potts, D.M. and Addenbrooke, T.I. (1996) The influence of an existing surface structure on the ground movements due to tunnelling, in Mair and Taylor (eds) 1996, 573–8.

Powell, D.B., Sigl, O. and Beveridge, J.P. (1997) Heathrow Express – design and performance of platform tunnels at Terminal 4, *Tunnelling '97*, Institute of Mining and Metallurgy, London, pp. 565–93.

Pugsley, A.G. (1966) *The Safety of Structures*, Arnold, London.

Rabcewicz, L.v. (1944) *Gebirgsdruck und Tunnelbau*, Springer-Verlag, Berlin.

Rabcewicz, L.v. (1964) The new Austrian tunnelling method, *Water Power*, **16**, 453–7, 511–14, **17**, 19–24.

Rabcewicz, L.v. (1969) Stability of tunnels under rock load, *Water Power*, **21**, 225–302.

Rabcewicz, L.v. and Golser, J. (1973) Principles of dimensioning the supporting system for the new Austrian tunnelling method, *Water Power*, **25**(3), 88–93.

Ray, K. (1998) Tunnels and infrastructure for metropolises: the 'habitat' prospect, *Gallerie*, **55**, 21–26 (also *Tribune* (ITA) June 1998, 23–6).

Reed, C.B. (1999) Choosing the right contract and the right contractor, in *Proc. Øresund Link Dredging and Reclamation Conf.*, Øresundskonsortiet, Copenhagen (eds C. Iverson and B. Morgenson), pp. 73–85.

Renato, E.B. and Karel, R. (1998) Tunnel design and construction in extremely difficult ground conditions, *Tunnel (Köln)*, **8/98**, 23–31.

Ridley, M. (1996) *The Origin of Virtue*, Viking (also Penguin 1997), New York.

Rimington, J.D. (1993) *Coping with Technological Risk: A 21st Century Problem*, The CSE Lecture. Royal Academy of Engineers, London.

Roach, M.J. (1998) The strengthening of Brunel's Thames tunnel, *Proc. Inst. Civ. Engrs Transp.*, **129**, 106–15.

Royal Society (1992) *Risk Analysis, Perception, Management*, Royal Society, London.

Sandström, G.E. (1963) *The History of Tunnelling*, Barrie and Rocklliff, London.

Sauer, G. (1986) The new Austrian tunnelling method; theory and practice, *Tunnel (Köln)*, **4/86**, 280–8.

Scott, P.A. (1952) A 75-inch diameter water main in tunnel: a new method of tunnelling in London clay, *Proc. Inst. Civ. Engrs*, **1**(1), 302–17.

Sharp, J.C., Endersbee, L.A. and Mellors, T.W. (1984) Design and observed performance of permanent cavern excavation in weak, bedded strata, in *Proc. ISRM Symp. Design and Performance of Underground Excavations*, Brown, E.T. and Hudson, J.A. (eds), British Geotechnical Society, London.

Sharp, J.C., Warren, C.D., Barton, N.R. and Muir Wood, R. (1996). Fundamental evaluations of the Chalk Marl for the prediction of UK marine tunnel stability and water inflows, in Harris *et al.* 1996, pp. 472–507.

Sharp, J.D. and Turner, M. (1989) Pipejacking through hazardous ground in St Helens, *Tunnels Tunnelling*, March, 49–50.

Shirlaw, J.N. (1996) Ground treatment for bored tunnels, in Mair and Taylor (eds) 1996, pp. 19–25.

Simîc, D. and Gittoes, G. (1996) Ground behaviour and potential damage to buildings caused by the construction of a large diameter tunnel for the Lisbon Metro, in Mair and Taylor (eds.) 1996, pp. 745–50.

Simms, F.W. (1944) *Practical Tunnelling*, Lockwood, London.

Singer, C., Horbury, E.J. and Hall, A.R. (1954) *History of Technology*, Oxford University Press, Oxford.

Skamris, M.K. and Flyvbjerg, B. (1996). Accuracy of traffic forecasts and cost estimates for large transportation projects, *Transport. Res. Rec. No. 1518*, National Research and Academic Press, Washington, DC.

Sloan, A. (1997) Grouting to the rescue, *Ground Engng*, August, 12–13.

Stack, B. (1995) *Encyclopaedia of Tunnelling, Mining and Drilling Equipment*. Muden, Hobart.

Sterling, R.L. (1990) *Legal and Administrative Issues in Underground Space Use*, International Tunnelling Association, Lyon.

Széchy, K. (1970) *The Art of Tunnelling*, Akadémiai Kiadó, Budapest

Tabor, E.H. (1908) The Rotherhithe Tunnel, *Min. Proc. Inst. Civ. Engrs*, **175**, 190–208.

Tait, J.C. and Høj, N.P. (1996) Storebaelt eastern railway tunnel: Dania tunnel boring machine fire – analysis and recovery, *Proc. Inst. Civ. Engrs Civ. Engng Supp.*, **114**(1), 40–8.

Talobre, J.A. (1976) *La Mechanique des Roches*, Dunod, Paris

Terris, A.K. and Morgan, H.D. (1961) New tunnels near Potters Bar in the Eastern Region of British Railways, *Proc. Inst. Civ. Engrs*, **18**, 289–304.

Terzaghi, K. (1961) Loads on tunnel supports, in *Rock Tunneling with Steel Supports*, K.V. Proctor and T.L. White (eds), Commercial Stamping and Shearing Co., Youngstown, Ohio.

Terzaghi, K and Peck, R.B. (1967) *Soil mechanics in Engineering Practice*, John Wiley, New York.

Thomas, H.R., Smith, G.R. and Ponderlick, R.M. (1992) Resolving contract disputes based on misrepresentation, *J. Constr. Engng. Man.*, ASCE, **118**(3), September, 472–87.

Thomson, J.C. (1993) *Pipejacking and Microtunnelling*, Blackie, London.

Troughton, V.M., Murray, L.V. and Murray, S.A. (1991). Prediction and control of groundwater, vibration and noise for construction of Hong Kong Bank seawater tunnel, *Tunnelling '91*, Institute of Mining and Metallurgy, London, pp. 411–23.

UK Nirex (1993). *Nirex Deep Repository Project, Scientific Update 1993*, Nirex Report No. 525, UK Nirex Ltd, Harwell, UK.

Varley, P.M. (1996) Seismic risk assessment and analysis, in Harris *et al.* (eds) 1996, pp. 194–216.

Varley, P.M., Darby, A. and Radcliffe, E. (1992) Geology, alignment and survey, *Proc. Inst. Civ. Engrs, Civ. Engng Channel Tunnel*, **1**, 43–54.

Vaughan, P.R., Kennard, R.M. and Greenwood, D.A. (1983) Squeeze grouting of stiff-fissured clay after a tunnel collapse, in *Proc. 8th European Conf. Soil Mechanics Foundation Engineering*, Vol. 1 (eds H.G. Rathmayer and K.H.O. Scari), Balkema, Rotterdam, pp. 171–176.

Vitanage, P.W. (1982) Importance of detailed morphotectonic and geological studies as a means of predicting potential hazards and problems in tunnelling and site investigations, in *Proc. IV Cong. International Association of Engineering Geology*, New Delhi, Vol. 4, Theme 2, Int. Soc. Eng. Geol. New Delhi, pp. 203–15.

Wagner, H. and Schulter, A. (1996) *Tunnel Boring Machines*, Balkema, Brookfield, VT.

Wallis, S. (1998a) London's JLE experience with closed-face soft-ground pressurised TBMs, *Tunnel (Köln)*, **2/98**, 15–29.

Wallis, S. (1998b) Sydney gas caverns, *World Tunnelling*, October, 393–7.

Ward, W.H. (1978). Ground support for tunnels in weak rock, *Géotechnique*, **28**(2), 133–71.

Warren, C.D., Varley, P.M. and Parkin, R. (1996) UK tunnels: Geotechnical monitoring and encountered conditions, in Harris *et al.* (eds) 1996, 219–43.

West, G., Carter, P.G., Dumbleton, M.J. and Lake, L.M. (1981) Site investigation for tunnels, *Int. J. Rock Mech. Min. Sci.*, **18**, 345–67.

Author Index

Subject Index